LOD

GAVIN KNIGHT

Gavin Knight's first book, *Hood Rat*, about gun and gang crime in the UK's cities, was shortlisted for the Orwell Prize and the Crime Writer's Association Non-fiction Dagger in 2012. To research it, he spent two years with criminals, frontline police units and gang members from the inner cities of Britain. His work has appeared in publications including *The Times*, *Guardian*, *Daily Telegraph*, *Prospect*, *Newsweek*, *New Statesman* and *Esquire*; and he has appeared on BBC, CNN, ITN, Channel Four News and Sky News. This is his second book.

'Wonderfully evocative'
Matt Lewis, author of *Last Man Off*

'Gavin Knight has managed to shine light where before . . . there was darkness, and that is no mean feat'
Literary Review

'As a cross-section of west Cornish lives, a celebration of brave eccentricity and a prose illustration of the way those lives overlap and interrelate, *The Swordfish and the Star* takes some beating'
Patrick Gale, *Guardian*

'A hugely refreshing dunk in the ocean . . . for pushing at the boundaries of what non-fiction writing can
~~~~~ ~~~me fascinating social
~~~~~ lauded'

D1336250

ALSO BY GAVIN KNIGHT

Hood Rat

GAVIN KNIGHT

The Swordfish and the Star

Life on Cornwall's Most
Treacherous Stretch of Coast

VINTAGE

1 3 5 7 9 10 8 6 4 2

Vintage
20 Vauxhall Bridge Road,
London SW1V 2SA

Vintage is part of the Penguin Random House group of companies
whose addresses can be found at global.penguinrandomhouse.com

Penguin
Random House
UK

Copyright © Gavin Knight 2016

Gavin Knight has asserted his right to be identified as the
author of this Work in accordance with the Copyright,
Designs and Patents Act 1988

First published by Vintage in 2017
First published by Chatto & Windus in 2016

penguin.co.uk/vintage

A CIP catalogue record for this book is available from the British Library

ISBN 9781784700997

Printed and bound by Clays Ltd, St Ives Plc

Penguin Random House is committed to a sustainable future
for our business, our readers and our planet. This book is made
from Forest Stewardship Council® certified paper.

MIX
Paper from
responsible sources
FSC® C018179

To Anna, Iris and Harvey

The road to the far west is a strange one, full of enchantments and dangers. At the end of the lean Cornish peninsula the land parts like a pair of hungry wolfish jaws, baring its teeth to the Atlantic. The top jaw is the Penwith peninsula. Penwith from its Cornish root 'penn-wydh' means 'the end of the end'. It is a remote frontier facing the primordial fury of Atlantic storms. The sea pounds black granite cliffs. Wind and rain flay the land without pity so shiny knuckles of rock protrude through treeless moor. People are at their most Cornish – defiant, a law unto themselves. This was the Cornish language's final outpost.

Across the bay the lower jawline is formed by the Lizard peninsula, which juts south into the Channel. It is Britain's most treacherous stretch of coast, notorious for wrecks. If a sudden squall blows up offshore, seafarers find themselves driven into the jaws. A fisherman casting his nets off this coast drifts over pirate ships on the seabed, their gold sovereigns preserved in silt. The holidaymaker passing through, clotted-cream ice cream in hand, sees the Lizard's quaintness: its tiny coves with whitewashed, thatched cottages and fishing boats pulled up on the shingle. But the 'quaint' surface is not the whole picture.

For two years I spoke to people who lived in West Penwith and the Lizard. They were kind enough to meet in person and be interviewed on tape. Some are fishermen, and all are connected to the sea. The stories of the men and women who live here map out a particular place at a particular time: the extreme west of the county, the communities of Newlyn, Mousehole, St Just, St Ives, Cadgwith and the Lizard, a huddle of grey roofs on the edge of the sea, at the beginning of the twenty-first century.

The fishing industry is in decline, but these are still the richest waters for white fish in Britain. Martin Ellis, a Cadgwith crabber known to his friends as Nutty Noah, is at the heart of the book. He is a classic Cornish character who stubbornly tries to revive an ancient fishing practice which many see as a lost cause. I allow my subjects to describe their lives in their own words and voices where I can, without any judgements or intrusion from me. Dialogue is taken from direct quotes or reported speech from interviews. These recollections are supplemented by research into published material which is not referenced except in the bibliography and acknowledgements.

I've been fascinated by Cornwall for a long time. I live a few hours away in Somerset, but my wife's family live on the Cornish coast, and I was married in a church in the dunes there, where John Betjeman is buried. I talked in person with a wide cast of seafarers and Cornish characters: from fishermen mending nets or drinking in the pubs along Newlyn quay, to a family who have lived in the ornate rooms of Trereife Manor since the eighteenth century; to Nutty Noah in his driftwood shack on the Lizard; to Billy Stevenson and his daughter Elizabeth in their houses on Chywoone Hill, looking down on the Newlyn fleet their family has built up over a hundred years.

Newlyn is the largest deep-sea fishing port in England. Like many villages, it is tribal. There are only eight or nine great fishermen who notice when and why the fish come in, and everyone else just follows them. These are men who can tell from the cloudiness and colour of the water, or from a change in the temperature, whether the plankton levels are on the rise. If plankton multiply, the mackerel will follow, and blue sharks will follow the mackerel.

Fishermen are a breed apart; they are hunters. If you can't find the fish, you can't turn a shilling. In the pubs on Newlyn quay – the Swordfish Inn, the Star Inn, the Fisherman's Arms – they are always pumping each other for information about who caught what: which fishing ground was it? What gurdies did he use? Or they are gossiping about some City moneybags who came down, bought a boat for £750,000, sprouted a hole, sank it and lost everything. These are the pubs that give the book its title: *The Swordfish and the Star*.

It is a dangerous life. Eddie the Ice died when a three-quarter-inch towing line snapped and took his head off with a noise like a bass string being plucked. You come into the Swordfish on a good day, when the boats have landed, and count how many fingers are missing. One guy, Fred Steel, lost all of his on his right hand, cut off at the knuckle, caught in a winch. The dole office told him that there were loads of jobs you could do with one hand. Another, Spooky, was on a beam trawler when the rope snagged on a rock. He threw himself to the deck, watched the rope fly back to the wheelhouse like a whip and smash the boat up.

The Swordfish Inn is a forbidding place that holds dark secrets. *The twenty-four-hour Swordy! You'll get killed in there!* Newlyn was like a Wild West bonanza town during the mackerel boom of the 1970s and early 80s. Fishermen were paid in cash, thronged at the bar. Some hauled in drugs that Spanish dealers had stowed in their lobster pots.

Debt-ridden fishermen are forced to go a hundred miles to fish off wrecks in winter gales. The life takes its toll. One shopkeeper in Penzance says fishermen come in wild-eyed, with the shakes. 'If you want a gun or a bag of heroin, ask a fisherman,' the local policeman jokes.

Over time many stories rise to the surface. A crabber sets sail for France with an aristocrat on board, who turns out to be a fugitive from justice. A painter who messed around with a fisherman's wife is found dead at the bottom of a tin mine. In winter, broke Cornish boys march on Mousehole to occupy the empty second homes. Squatters live on the forgotten cruisers of faraway City traders. Through these linked stories, the remote west of Cornwall becomes a microcosm for modern Britain, with characters from all levels of society. The coast is treacherous with its hidden and unpredictable dangers, but there is also treachery of another kind: human betrayal and deception.

Cornwall was once a separate country with its own language, culture and folklore. The far west has been the kernel of rebellion for centuries. Artists are lured here by the drama and beauty of the landscape, but instead encounter unquiet forces at work.

Dark secrets lie buried in the land beneath the ancient burial sites with lichen-covered standing stones.

We associate Cornwall with holidays by the sea, with myths and legends. It's the end of the railway line. But the place we escape to is often the place where we have to confront things about ourselves – both individually and collectively. Above all, Cornwall is the place where people find the unexpected.

A ship passes through the waves, and the water closes up behind, leaving no trace.

Nutty Noah is six feet four, with wild reddish hair and a tufty beard. He has a hopeless, Micawberish optimism. If he hauled in a large catch he'd sprawl headlong in it on deck and pretend to swim, in his dirty Guernsey and oilskins up to his armpits. Nutty tried his luck at anything – whelks, pilchards, conger, even sharks.

Nutty's real name is Martin Ellis. Born in 1954, he grew up not far from Newlyn in the sheltered fishing cove of Cadgwith, on Cornwall's southernmost point, the Lizard peninsula. His dad, granddad and male ancestors going back five generations had all been fishermen. Right up to the nineteenth century this part of Cornwall was known for its smugglers and pirates, men familiar with tiny coves and secret, moonlit beaches like Cadgwith, and who fuelled the black economy. It had been a fishing village since medieval times, with its thatched cottages and whitewashed walls. Today only a handful of crabbing boats still launch from the quay.

When he was sixteen he played in the punts with his teenage cousin Phil. They would just muck about, it wasn't proper fishing. Punts are small, open wooden boats used for inshore fishing, powered by oars or a motor. One day Phil's dad, Uncle Harry, took him out crabbing. The heavy withy crab pots were stacked in a row and flew off the back, sinking down to the seabed. The pots were twisted and threaded from bendy willow cut from local beds. There was loads of rope on deck; it writhed and slithered over the side as the pots went off the back. Rope everywhere. Uncle Harry was busy looking out for his son Phil and did not notice his foot resting on a coil while it was playing out. The coil snapped around his ankle, wrenched him clean over the side. The weight of pots and drag of the boat hauled him down so fast he never came up. Very few who are pulled below ever do. Uncle Harry drowned off the Lizard with a rope round his leg. Phil brought the boat back in, raised the alarm then set back out to sea in it again to find his dad. They called him back. Out of his parents' council house window, Martin saw Aunty Nora coming up the path, wild-eyed, her face puffy and crumpled, pulling her cardie tight around her neck. A local diver went down to the seabed and found the body. Phil and his family all moved away after that.

At nineteen, Martin went on board a trawler to have a cup of tea with a local fisherman, Plugger, who invited him to join them at sea. Summertime fishing meant catching crabs and lobsters. Winter meant mackerel. He loved the idea of following in his ancestors' footsteps. For a young man it was exciting to steam off in the dark, the hour before first light, and see as many as 200 boats stretching out to the horizon as the sun came up. Men came down from Looe, from Mevagissey, Newlyn fishermen came up east of the Lizard. Even Russian trawlers anchored up. First and last light saw the best of the mackerel fishing.

In the 1970s, hand-lining mackerel gave a living to thousands of Cornish people, but it was brutal, tiring work. Martin struggled to stand up while the boat rolled on a winter sea. He'd use coloured feathers or plastic slotted on the shanks. He worked twenty-five hooks that bloody hurt if they caught him. He'd go home with swollen, pink hands when that happened. Mackerel aren't intelligent fish. If you found the shoals, they would bite the hooks readily enough, but they would fight hard to get away, dragging the line back through the water. He would hoist the individual lines and slam each fish against the side of the hull until its jawbone broke and it dropped onto deck. Then he stowed them and took up the slack on the lines. He had to do that fast: in fifteen seconds a slack reel could tangle up into a bird's nest. Then you'd have to cut it loose and the skipper would be well pissed off that you had lost him ten stone in fish.

That was how to catch mackerel in those days, before the bigger fisheries arrived in 1976, before their high-tech trawlers, purse-seiners and beamers steamrolled the seabed into clouds of coral dust and dead fish. Trawlers drag nets along the seabed. In beam trawling, the mouth of the trawl is held open by twelve-metre metal stanchions, as long as London buses. Purse-seiners surround an entire shoal with deep curtains of net, with floats at the top, weights that sink at the bottom. Once surrounded the net is then closed at the bottom like a purse, trapping the fish.

Friday nights he blew his earnings in the Cadgwith Cove Inn. Fifty fishermen with their girlfriends, crammed into a small bar, roaring out shanties like 'Row Boatman Row', 'White Rose' or the Cadgwith anthem, 'Robber's Retreat':

Come fill up your glasses and let us be merry,
For to rob bags of plunder it is our intent.

Martin's deep baritone rang out. On the swell of half-drained pints of Atlantic ale, the singers swayed in song, fingers closed around loops of rope hung from low, black beams. Light from the bar glinted off a brass barometer, captain's clock and photos of Cadgwith fishermen. They sang the old heart songs: 'Lamorna', about a man who rides all night in a moonlit cab next to a maiden with a dark, roving eye and rain-slaked hair. He lifts her veil to find it's his wife. They belted out the defiant Cornish anthem:

I've stood on Cape Cornwall in the sun's evening glow,
On Chywoone Hill at Newlyn to watch the fishing fleets go,
Watched the sheave wheels at Geevor as they spun around
And heard the men singing as they go underground.
And no one will ever move me from this land
Until the Lord calls me to sit at his hand.

Then it was off to the net loft for rum and beer after hours. They drank rum with shrub, a fruit syrup which Cadgwith smugglers used to mask the briny, salty taste of seawater that seeped into barrels sunk and hidden where no excise man could find them. One particularly successful week Martin and his dad threw a legendary piss-up that went on for two days.

Martin liked the freedom of being outdoors. He thought about North American Indians, catching salmon in traps, hunting deer with a bow and arrow. Martin's version was seeking out a shoal fathoms down with an echo-sounder. Mackerel like a certain depth. Forty fathoms below the surface he had seen shoals of fish which measured twenty fathoms deep: the size of a hotel in volume, hundreds of tons of fish. At nights he'd go home to the little house in Cadgwith he shared with his parents.

You're told early: it's in Cornish blood to be good fishermen, to be genuine in your work whether you're building a hedge or digging a hole. It's the heritage. These are the original Britons. The Romans never made it west of Exeter and the Devil was too scared to cross the Tamar.

After his father had been made redundant from nearby Culdrose naval station (the biggest helicopter base in Europe), they fished together. Their eighteen-feet fibreglass boat was called the *Tidos*, which was 'sod it' backwards, the name they also gave to their two-day party. In 1974, when he was twenty, Martin took out a bank loan to buy the *Samantha Rose*, a wooden, twenty-feet crabbing boat. Like many local boats it was designed to fish for mackerel in the winter, then switch to crabbing and lobsters in the summer. Martin raised the mizzen sail to steady her. Lobsters and crabs lived amongst the rocks. It demanded fine seamanship to steer close to granite cliffs and inaccessible coves. 'The closer to the bone, the sweeter the meat,' Plugger used to say. On the way back Martin would take hen crabs straight off the cove. He dropped a string of lobster pots over the side to the seabed, a dozen at a time, and marked each one with a buoy on a rope. The lobsters crawled in a small entrance after the bait and got trapped inside. It meant the boat had pots stacked up and tons of rope on deck. Rope everywhere. He thought about Uncle Harry pulled over the side, dragged down by the weight of pots and pull of the boat. The tides are strong off the Lizard when they go against you, drawing you onto the rocks, smashing up hundreds of pounds' worth of gear each grinding second. A white mist rolls in and swallows up the boats. Mast and wheelhouse fade into a ghostly outline, traced in watery strokes, then nothing can be seen. Only shouts and clanks. A huge brush paints over the horizon, town, cliffs and warning buoys until a vast wall of white is all that's left. Then the known world ends at the bowsprit. Only the boat exists.

The same as it was for his ancestors. He knew he was doing one of the hardest and most dangerous jobs in the world. He had to stand on a platform which was rolling as well as the bow going up and down, while hurling out heavy pots with a railway shoe, or twack, as an end weight for each string of pots. He hauled in the pots, bailed out loads of small crab, took the large crab and stored them in another pot. Each pot stowed in sequence for next time. Some people cut a hole in the back of the boat and shot them out the back end. Brilliant idea! It was a darn sight easier than picking them all up and hauling them over the side again. Each time he heaved a pot over he had to stand behind the rope or end up dead. One man stepped into a loop of rope, it clamped tight around his ankle like a snare, yanked him ten feet to the stern then ripped his foot clean off. Martin still sees him limping through Helston with his false foot.

After he'd bought the *Samantha Rose*, Martin had to stump up for lots of kit: a decent hauler to wind the net in; and a derrick – a strong crane arm – which took the strain against the tide. Floating orange buffs to keep the gear up, chains. Plus bait. It all cost money. Mackerel nets floated near the surface. Spring tides were the worst: the full moon pulled the waters up so high that the nets were drawn out flat and mackerel didn't like it. Those nights, the pubs filled up with netters.

As good landings of handline-caught mackerel came in, the fishermen were paid in cash by their skipper in the Swordfish. There was good money to be made. For the first time in years, young men like Martin were following in their fathers' footsteps and making a living out of fishing.

But rumours of the mackerel shoals in Mount's Bay were slowly spreading outside Cornwall. One evening the Cadgwith fishermen were heading in when they spotted a dark ship nearby, over seventy feet long, looming over them. It was rusted like a coaster, low in the water. Merchant ships freighted oil or coal in bulk around the British coastline; they had low hulls so they didn't snag on the dangerous reefs. But this ship wasn't going anywhere. It wasn't a coaster. It was a vast fishing vessel, here for their mackerel. In the Falmouth pubs there were gravelly, hoarse Scottish voices calling for pints. Sandy-haired Norwegians, with eyes glaucous like boiled pollock, were telling yarns of how they read the sea, searched the remote Arctic Ocean, how they pushed themselves to the limit.

'In Norway we have custom called *hal* – lots of sex night before means good catch next day.'

They laughed. One wore what looked like a white letter J, but on a closer look it was a silver fish hook, the ancient symbol of Batsfjord, where he fished the deep waters of the Barents Sea. The Norwegians drank shots and dark bitter, joking how back home amber aquavit helped the *lutefisk* swim down to their stomachs. The Scots tipped pints down, paid for with wads of cash. These invaders came from remote places like Fraserburgh or Lerwick. They had fished their own North Sea herring shoals out of the water with monster ships: giant purse seiners and pair trawlers. With the herring fishery destroyed the ships had been slumbering in harbour with nothing to do. So the Scots headed down south and could not believe the teeming shoals of Cornish mackerel waiting for them in Mount's Bay. First there were three boats, then there were forty-five.

The mackerel boom turned Newlyn into a bonanza town. Under the street lights, they were loading lorries to be shipped out round the clock. One night in the Swordfish a fight broke out between East Coast and West Coast Scots. East and West Scots fighting meant it was about religion, Catholic and Protestants. It was a massive ruckus with five crews all fighting. You rarely see a policeman in Newlyn but that night they came with live ammunition and cordoned off the street. One of Newlyn's hardest Scots, Fish, dived in and swam across to the new harbour. He swam to get to them to carry on fighting. Soon there were more swimming across the harbour. The whole of Newlyn was swarming with coppers. The fishermen broke each other's boats up. It was fucking chaos. They put deck hoses, seawater hoses, on in the fish room, trying to sink the boats. One man was stabbed, slit from the wrist to the elbow, but they couldn't find the evidence. In those days everyone carried a knife. Once one used his, the others joined in. They were all trained to use them, to gut the fish. Now they are not supposed to carry one if it's over three inches.

Another time, years before, a row had broken out in the Swordfish, the police came and the fishermen told them to fuck off. They threw one copper through the window. They were a rough bunch of lads in those days.

The Swordfish, Star and Dolphin are in a triangle, a few yards apart: it was called the Bermuda triangle because fishermen would disappear into it, never to be seen again. Many of the long-standing fishing families, with strong religious, teetotal backgrounds, would not be seen dead and gutted in the Swordfish. In the front of the Swordfish people would be drinking but in the back, there was any drug you wanted. One time, to get to the bar you had to step over a fisherman lying on the floor unconscious from coke. An outsider walked in and saw a fisherman's wife knock her husband out for chatting up another woman. She laid him clean out.

A lot of bikers from the Penzance chapter of Scorpio, based in the Seven Stars, came down to Newlyn. As legend has it, the Newlyn fishermen outdrunk, outdrugged and outfought the bikers. One fisherman called 'Warrior', six feet four and with a wild long mane of hair, roared a motorbike straight through the Swordfish. Another sent three of them flying through the door of the Star. Today, Squeeze, an ex-biker with a sleeve of tattoos, keeps the cellar stocked in the Star at 10 a.m.

In the mackerel boom there was real tension between the Scottish invaders and the Cornish fleet. After a while most of the Scots didn't risk going into the Newlyn pubs. The skippers thronged into an Italian restaurant, La Cucina, in Falmouth, near Custom House Quay. During the day they tuned their VHF frequencies until they could eavesdrop on the local handliners out searching for the shoals. When the Cornish boats found a shoal, the Scottish ships encircled them, swallowing the shoals up until their vast holds were swollen night after night with thousands of tons of mackerel which were pumped ashore in Falmouth. They landed more in one day than the local boats caught in a month.

Other giant stern trawlers came down from Hull and Grimsby; these industrial ships hauled their trawl up ramps fitted at the stern. They had decimated Arctic cod stocks and were banned in the Cod Wars of the 1970s. They in turn loaded their catch up onto Russian factory ships on anchor at Carrick Roads, the natural Ice Age harbour near Falmouth. The Scottish pair trawlers hunted in groups, keeping in touch by radio. One shot the trawl, the second took the other end of the net and they dragged it together until it was full. Meanwhile the third boat hunted for the next shoal. Then as one boat hauled the catch onboard the other two started another trawl. They did not waste a second. When they caught more than they could hold, sinking low in the water, they spewed dead fish out until the whole of the south Cornish seabed was carpeted with rotting, stinking mackerel carcases. Dolphins and pilot whales were also killed.

'How much can your ship load?'

'250 tons of mackerel.'

'We don't want you in our Cornish waters.'

'I'm a British-registered vessel.'

These monster ships conquered Cornish waters. It was like the D-Day landings in Falmouth, with sightings of thirty-three ships from seven different countries. St Ives, once a quaint fishing town, became a prosperous harbour again: fleets of eleven massive Norwegian or Dutch pursers sheltered in the bay. Giant freezer factory trawlers gobbled up fifty tons of mackerel. The locals watched these invaders.

'They don't know these waters,' they said.

The Cornish have a legend about a lady in a flowing white dress with a rose between her teeth who stands in a hollow of the cliffs near Land's End. If you see her, you die. Just as ominous is the sight of the white mist that rolls in to catch seafarers. The mist cloaked a Romanian factory trawler, *Rarau*, fishing off the Scillies. They had never seen fog that thick. But they kept fishing. The crew spotted the telltale white foam threshing on the surface of the water: it looked like a shoal of mackerel. They closed in and tried to trawl it up. It wasn't a shoal threshing; it was the waves breaking over a black granite reef, the notorious Seven Stones. This rock is so dangerous it has claimed over 200 wrecks over the centuries. Wedged down in the darkness of the depths lay the wreck of the world's largest supertanker, the *Torrey Canyon*, gone down only ten years before. As the mist swirled by they saw the Seven Stones reef rearing up out of deep water. It is nearly two miles long. The Romanian ship had an enormous crew of eighty-four. Some 2,500 tons of factory ship struck the north rock with a grinding scream of twisted metal. Seawater roared into the engine room and hold. It was speared on the reef. Another Romanian factory ship rescued the crew and took them back to Falmouth. A week later the *Rarau* slipped off the reef and was lost in deep water.

Martin knew the shoals were vast for 150–200 boats to be out there mackereling, trawlers coming in and out, the purse-seiners and Russians. But prices were being hammered down to forty pence a stone. The handliners in their small wooden boats could no longer afford to miss a day's fishing and had to risk their lives going out in winter gales. The industrial ships were fine in poor weather; they were powerful seventy-feet steel-built craft.

The Cornish are fierce, proud fishermen. They protested any way they could. One Cornish fleet of handliners was working a shoal off Dodman's Point near Mevagissey when a Scottish pair encircled them and started to trawl through the shoal. The fleet decided to fight back and broke off fishing. Twenty small boats tucked themselves under the steel-built bow of the seventy-feet trawlers, forcing them to slow down so their nets sank and snagged on rocks.

'If we see you onshore, you're dead men!' the Cornish boats yelled.

After hollering and waving at the small boats, the Scots' nerve broke and they hauled their nets and sailed away.

Martin took a day off mackereling and decided to look for other fish. He had seen grey mullet on top of the water in summer, but they were too shy and quick for anyone to catch them. They were smart fish.

'If you can see mullet on top of the water now, you'll see them in winter,' Buller had told him. Buller was a man you listened to in Cadgwith. He was born in 1901. He kept the old ways alive down the cove. He revived pilot-gig racing, where six men rowed a 32-feet boat built of Cornish leaf elm through the sea. It had a beam of four feet ten, and was one of the first lifeboats – it had rescued people as far back as the seventeenth century. Buller was a lovely man. He sat on the cove's wooden bench, known as the Stick, and often sang there in powerful, stirring harmonies with his friend Harvey Tripp. Martin loved listening to his tales of the old days. Buller was happy to talk. One day he told him something that nearly made Martin fall off the bench.

When the bleak Cornish winter began to bite, the hungry villagers would line the cliff top, squinting out to look for gannets diving, for changes in the colour at the edge of the water. He told Martin how, in the last century, lookouts would cry out 'Hevva!' when they saw a shoal and direct two punts below with arm signals, using torn gorse bushes, so that they could encircle it.

'In Cornish history there are few sights as spectacular as when the shoals of pilchards were caught with large nets like that. All those men working together to find and process the fish. Your great-granddad used to look for shoals from the top of the cliffs. He was a huer.'

The seine boat would aim to enclose the entire shoal in a wall of net. The net was a monster. He explained how they winched up the bottom of the net, using rings of rope to draw it tight like an old-fashioned leather purse. This practice had died out in Cornish fishing at the turn of the century.

On shore a team of blowsers would march round the capstans, hauling the net full of fish into shallow waters where it was moored up. The sea's surface would break with a teeming, dazzling mass of silver fish. Pairs of men in oilskins and seaboots filled up the small boats by dipping baskets in the fish until, after three days, there was nothing left. This was called tucking the seine. By nightfall the tuckers were covered in silvery scales.

'We used to catch great shoals of grey mullet,' added Sharky, another older fisherman, related to Martin. 'You could see them when they were close together as colour in the water.'

'Like mackerel lashing on the surface?'

'No, these were down in the water. You would have to stay in one place and keep watch from the cliff for a long time.'

This was proper hunting. Martin had never heard of the technique before. The following day, he wrapped up warm and walked around the high cliffs of the Lizard looking for shoals like his great-granddad did.

These cliffs erupted 400 million years ago in green and purple serpentine rock, rugged and scaly. He could hear the pebble surf rattle far below as he peered out over tussocks and springy gorse to the grey sea surface.

The Cornish word for the colour of the sea is '*glas*'. It can mean green, grey or blue. If you are a fisherman in Cornwall the most important thing is the state of the sea, rather than its colour. What's the colour of the sea today? It's *glas*. It's always *glas*. *Glas* is a wry, playful word that predates Isaac Newton and his spectrum of light. It is an important clue about the Cornish not wanting to be pinned down. They have another word – *gwerth* – for green things that aren't living: a green bottle, Barbour coat, or wellington boot. Only living, organic things like the sea are '*glas*'.

Martin could see seals and choughs up there but only a trained eye could see the grey mullet when they were bunched together. He looked out from Kennack Sands and Pentreath, west of the Lizard and out from Gunwalloe, where a church is set below the cliff. After a long time standing in the cold, fingers going numb, he made out a browny, reddish stain near the surface. It looked like a shoal, but as he kept looking at it he realised it wasn't moving. It was only the sea cracking over hard ground or sand or rocks. Sometimes it moved only to lift up into a scud of spray, a mass of vapour clouds, driven by the wind.

'You have to look away,' the old fisherman told him. 'Bring your eyes back to see if the shoal has moved.'

It was best in winter when wind blew from the north over a calm sea and the fish were shoaled up; they went into hibernation mode then. Most people in the cove thought Martin was crazy going up on the cliff to look for shoals of mullet. Why wasn't he out catching mackerel? So he spent several years scratching his head, wondering. A lot of people didn't think anyone could still spot fish from up there. They gave him a new nickname – Nutty Noah.

'It's not possible,' they said down the cove, shaking their heads. 'Mullet hasn't been caught like that for fifty, sixty years or more.'

Fuck it. When the winter storms came, when it was too rough to go mackereling and he had a day off, Martin wrapped up warm and walked up the cliff around Coverack or Kynance. One day he made out another browny, reddish stain near the surface. He looked away and looked back. This time he had got his eye in and he found he could follow it through the swell, see it rippling with the motion of the water. He punched the air as he watched it move slowly through the water. He sang all the way home. Now he needed to work out how you took grey mullet like that.

Back in Cadgwith they thought he was joking.

'You need ring nets to catch shoals like that,' they laughed. 'No one's used ring nets in donkey's years.' Not many people were interested; they didn't believe he'd seen a shoal. He persuaded a handful of old crabbers, Uncle Arthur, Sharky and Philip Burgess, to go to the cliff with him and have a look. They realised it was a really big shoal. Word spread. Soon everybody came out to take a look. They waited for three days with two sixteen-feet punts and a huge monster net ready. Sharky and Uncle Arthur stayed on the cliff where they waved and shouted themselves hoarse, directing the punts around the shoal until they overlapped.

Then the men in the boats beat the water and frightened the fish into the net. It was a good catch, nearly 670 stone. A huge shoal worth £300.

There was a brilliant atmosphere down the cove when the shoal came in. Different families came down and helped take the fish out of the seine net. The fish were trapped in the mesh by their gills and it took a lot of manhandling to pull them out. They worked over the depth and length of the net: ten fathoms deep and a hundred fathoms long. It took a while. They ordered the lorry, put the fish in boxes to be sent away. Women came down with cups of coffee or pints of beer.

From that day they started talking about Martin as a pioneer: a fisherman to watch. They fished that way once or twice a year, when there were no mackerel to be had. But sometimes the catch was divided into as many as eighteen shares.

When the winter mackerel season was over, he kept crabbing down the cove. There was a danger that a scalloper might drag through his gear and tow it all away. Scallopers dredged a bar of eight-inch steel teeth along the seabed which dug out the scallops. As he lurched and swayed on the bobbing platform fierce winds and tides dragged him closer to sheer granite rock faces, where the restless sea seethed into gullies and dark cracks. Only the tumbling acrobat, the red-beaked, red-footed chough, went closer. He smiled up at the birds. They are the crows of Cornwall. Legend has it that the soul of King Arthur departed this world in the form of a chough, its red feet and bill signifying Arthur's bloody end. They take their place on the Cornish coat of arms next to the tin miner and fisherman.

In Cadgwith Cove the fishermen started the day on a tractor, which shoved the crabbing boats over the steep, wet shingle into the sea. It took away the spine-cracking effort of hauling heavy boats into the water. So it was no surprise to Martin to see a tractor one day rumbling through the cove. But this one made him stop working. He halted with his heavy tea chest of crabs. The tractor had a link box mounted on the back, a five-feet metal box used for transporting logs or sheep. In the link box stood a strikingly beautiful girl with long brown hair. Martin had never seen her before.

He asked around. His brother André knew her from the local drama group. Her name was Sally. She was married to a local farmer who spent his time looking after his herd of Friesians. She was a free spirit. On horseback she trekked along miles of coastal bridle paths, went on hacks and rode over beaches.

Sally had grown up in the suburbs of south-east London, in a little end-of-terrace house in Sidcup. At the weekends, in a haze of dragonflies, she watched her father fly model airplanes or steer model boats around the ponds of Dartford Heath. She was nearly an heiress. Her immensely rich grandfather visited in gleaming Daimlers or Bristols bearing lavish presents. He was the boss of the family's fourth-generation cardboard-box business. When he died Sally's dad became the factory boss. But the grandfather had married a wife much younger than Sally's father; she moved to Australia with the fortune. Sally grew up with London's Underground, escalators, the zoo and museums, but she felt happiest on family holidays in Cornwall, being flung against the ropes in the back of her dad's short-wheelbase Land Rover as it slid around off-road. She became an outdoors girl who enjoyed cutting cauliflowers in the rain. She remembered the first time they drove down the steep, winding hill to Cadgwith Cove. In her mid-20s, after years sitting in airless London offices, she and her sister married two brothers and moved to the Lizard.

At Cadgwith Cove Inn her friends introduced her to Martin, an up-and-coming fisherman. He was always friendly and pleased to see her. Sally could see he was at the centre of life in the cove. The younger fishermen looked up to him. He towered above the singers in the inn and drummed his hands hard against the black beams during the chorus of 'South Australia', which he sang in his distinctive singing voice.

Farmers waited through the year for their yield to grow. The fishermen got their yield in one day, harvested from below the waves. It was a gamble that all depended on the weather. Sally found it exciting. There was a whole romance to fishing that didn't exist with farming. After three years she split up with her husband. He moved out with another woman and left her to prop up his father, in a bungalow on the hill in St Ruan. She was happy about it. Her friends said it was the best thing that could have happened.

One night there was a terrific storm down in the cove. Sally went down the next day to see what the beach looked like. She walked backwards up the hill squeezing the button on her little Instamatic camera. She framed the debris, hurled around by the violent wind. Then she heard a voice behind her.

'Coo, look at her taking photographs.'

She turned round. It was Martin with another fisherman. He nodded at her little Jack Russell, Shilling.

'Is she good at hunting?'

'You bet.'

'Fancy a walk out to Soapy Cove to see if she'll catch a rabbit?'

Shilling disappeared down a rabbit hole and popped up twenty yards away, with no quarry. The Cornish heather was springy underfoot. Bog and heath grew wild around them, purple musk thistle swayed in the wind. They laughed as they scrambled up rocks, up steep scree to regain a grassy track. They caught their breath and looked down at the ruin of Jollytown, between Mullion and the Lizard. From the clifftop they looked out for dolphins and basking sharks. Hardy Highland cattle peered at them through their overgrown ginger fringes, lowering their horns to graze the bracken. They climbed down to Soap Rock, the quarry for soapstone and porcelain. Then took a steep path back up again to the clifftop. The cliffs had collapsed as if a giant had fallen down; the slope of boulders led them down to the turquoise water and white sands of one of the prettiest beaches on the Lizard, Kynance Cove. That night, as it grew dark, they chatted on the Stick, the bench where Buller and Harvey sang the old heart songs, where all the fishermen gathered, gossiped and put the world to rights. Martin cradled her beautiful, young face in the moonlight and kissed her.

Martin could see the sea excited her, so he took her out in the *Samantha Rose*. She thought it was special: wooden, painted red and white; such a lovely boat that it called for people to go out on her. The boat was small enough to get in the little coves, where the sea seethed and boomed. They could moor off and go to Lankidden, a secret, secluded beach reachable only by coastal path. Martin handled her skillfully, so they went close to the Devil's Frying Pan, where the roof of a sea cave had collapsed leaving a 200-feet hole that churned and boiled in rough weather. They even went right up to the mouths of caves. As a courtship of a girl from a London suburb, it was just magical. She shared Martin's excitement for his plans. She saw him as a pioneer and noticed how other fishermen respected him.

He took her up on the cliffs looking for mullet.

'Look! Look, see that bit there,' he said, pointing. 'See how it shines.' If she looked through the water she could just get a glimpse of the mullet. She gasped and dug into the tufts with her nails. They were almost hanging over the cliff edge.

They courted through the autumn, when the tourist crowds had gone. On Christmas morning Martin told Sally he loved her. Later he dropped her off at work, at Dr Cuff's, the local doctor.

'I've got a bit of money. I'd like to put a ring on your finger.'

She watched him go then turned to Mrs Cuff.

'I've just had a very funny sort of proposal, Mrs Cuff.'

Sally was not sure. Could she go and commit herself, after she'd just left behind a three-year marriage to the Friesian farmer? But she'd been a young girl and it had not been a good relationship, whereas she and Martin got on really well.

'You can see by the way he looks at you, he's totally in love with you,' Mrs Cuff smiled. She had four children, so she knew what it was like. That afternoon Sally went with Martin to Falmouth to buy a rope, then back to Dr Cuff's house. Martin stood in the doctor's doorway.

'Come here a minute.'

Sally came over and he squeezed a ring over her knuckle. Sally walked in to another room, where Barbara and Peter Cuff were.

'We're engaged!'

Out came the sherry. They went down the pub. Everybody was delighted.

'You are such a well-suited couple.'

Sally was as daft as Martin. They were engaged in January 1984. Even though her father loved chatting to the fishermen down the cove he told her he didn't think there would be any money in fishing. Martin was a young fisherman of 29 with big ideas. She idolised him and went along with what he said. Some of his big ideas she'd live to regret.

Martin only let his mum and dad know after it had happened. He just walked in and said, 'Hi Mum and Dad, we're engaged.'

Over that winter the sea was so choppy it was difficult to launch from Cadgwith, so they moored up at Helford, toing and froing between the two. Sally loved the Helford River. It was completely unspoilt. In its deeply sheltered valleys open fields and ancient oak forests ran down to a rocky shore dotted with little beaches and hidden creeks. Further north on the Fal River the two sides were linked by a clanking chain-pulled ferry that dated back to the Middle Ages. It was so far from the derelict wharves and warehouses on the Thames by her dad's main factory in Bermondsey.

The outboard whined as the little punt took them out to the mooring. Sally glanced at her finger. Set in the ring was a violet amethyst. If he hadn't had to buy the rope in Falmouth she would have had a diamond.

In August, Sally was heaved unceremoniously in her wedding frock onto the *Samantha Rose*, where they cut the cake – two tiers with the Lizard lighthouse. Some friends did up Martin's loft down the cove for the reception. They'd got in 200 pints of Blue Anchor beer, a sweet Cornish bitter from the 500-year-old Helston inn where tin miners picked up their wages, and 200 bottles of wine. They took photos. Had a church blessing. Rather than have presents they asked people to pay for themselves at a little restaurant in Coverack.

Someone suggested they shove a caravan in Martin's uncle's garden in Prazegooth, on the other side of Cadgwith. Her father lent them £1,000 to buy a big one. It was towed right next to the family bungalow where Martin's mother and her thirteen brothers and sisters had all been brought up. They set it down beside two sheds and Martin built a connecting roof. Sally was very happy to be so close to his family. She'd lived with a Cornish family for three years and felt she knew the Cornish well. She brought a bit of new blood into the family, otherwise Martin might have ended up marrying his cousin Luke.

Those days in the caravan were brilliant. They lived there with seven cats, which was probably six too many. Martin loved being out in the elements. They could hear the rain on the roof. Sally used to have a caravan growing up, on the edge of Bodmin where her father had opened up a new box factory.

It didn't take long to start a family. They had two girls.

Those early summers together were wonderful. Her nan had given her a Silver Cross pram and she took the babies out. By then she had an English setter, a dog well-suited to families, which trotted along beside. They walked down the hill to the cove every day to meet Martin from fishing. They'd sit and wait to catch a glimpse of his boat as it came round the rocks. They were always pleased to see him. He was usually crabbing, carrying the crabs up the beach in a 125-pound tea chest on his shoulders with water dripping down his back. After a hard day's work he'd walk up the beach to the lorry. Then they brought the tractor in. The winch brought the boats up. The crabs were carried in bongos, like washing baskets, and put into the link box on the back of the tractor. Martin was quite good, he was always back for night-time.

Come November it was cold for toddlers in the caravan, and their little gas heater clagged the windows up with condensation. It smelt of damp. The only really warm spot was by the wood-burner in the lounge. They weren't able to buy any meat, sausages, or bacon. They couldn't eat the catch. So Martin set out with his twelve-bore. He was always doing something, if he wasn't down the cove he was off shooting, which Sally never minded because she knew he was going to come back with some food. All the meat they had, he shot. The secretive woodcock was in season in the winter on the Lizard. Their daughters were raised on pheasant and woodcock. Pigeons. Rabbits. Martin plucked them quickly. Sally roasted the pheasant, casseroled the pigeon and sweet, gamey woodcock on slow, then heated up the leftovers the next day. Made it last longer than one meal. The kids didn't complain. They didn't know.

They lived hand to mouth. They owed so much money all the time. Every time he had a big catch, most of the money went on paying the diesel bill and breakdown bills for the boat. Right the way through money was a worry. Sally found she could never budget. She had to wait to see what came from beneath the waves. When he caught a lot of fish they would all go to the big Safeway in Redruth, the nearest supermarket in those days. They piled the basket up with tins of soup, beans and a few luxuries. The most luxurious thing they bought, but only when the kids were older, was some Chinese food Martin fancied. You chose what you wanted and tucked in.

The winters were hard for Sally while Martin was off fishing. She couldn't get out much and it was claustrophobic with both kids in the caravan. The girls shared a tiny bedroom. Martin once made up 200 pots in the lounge, in the warmth of the wood burner. The girls loved it. They used to have games in them. An ex-fisherman came by, looked at all the gear spread around and said: 'My wife would divorce me for less.'

Sally listened to the radio 24/7 and made curtains. She knew the other mums by sight, but hadn't got to know them terribly well. It was a little bit lonely, but they called in on Martin's mum and dad, who lived a few streets away, nearly every day. Toots, their second daughter, cried a lot. Once she started at 3 a.m. Sally got up, changed her nappy. Martin woke up. Sally picked up the nearest thing, which happened to be cotton wool balls, and started trying to stuff them in the baby's mouth.

'I don't think that's a very good idea,' Martin said.

The girls learnt how to swear like a fisherman.

'You mustn't say that, it's a fisherman's word,' Sally said.

'Not all fishermen swear you know,' another mum within earshot said. Most of the girls' friends were fishermen's children, a real fishing community. They were forever on the boat with Dad, going round the block. Because the children were brought up like that, it wasn't terribly special to them. But it was special to Sally, being an outsider, looking in. They didn't know how lucky they were being able to go down to the beach every day, have cliff-top walks.

Someone reported them to the council for living in a caravan, so after a couple of years, in 1987, they moved into an ex-council house in Ruan Minor, which Sally bought with money she inherited from her father when he died. Most fishermen lived on top of the hill, in Ruan Minor; the quaint old fishermen's cottages down in the cove had been sold to incomers.

'They want quaint little places and quaint people. Whatever quaint is,' Martin said. 'I suppose I'm fucking quaint.'

Martin was deeply proud to be Cornish, from the Lizard. They'd always fought to keep the old ways alive. It was in the sixteenth century, on the Lizard in Helston, that a mob dragged the archbishop's man outside and stabbed him. The archbishop was trying to wipe the Cornish language out. He made everyone in church use a new prayer book, in English. No one spoke English; they didn't understand it. They were outraged. The ancient Cornish tongue, Kernewek, was related to Breton and Welsh and was spoken throughout the ninety miles of the rugged peninsula. They marched across the Tamar and besieged Exeter. They were fierce fighters; their longbow archers were deadly. So the king hired German mercenaries in flamboyant puffed and slashed doublets; 900 Cornish prisoners had their throats cut in ten minutes. Thousands more were drowned in the river at Clyst Heath. A Cornish priest was hanged from his own church tower.

The Cornish language clung on in the Lizard, in the remote cliffs of West Penwith. Landewednack, two miles from Nutty

Noah's house in Cadgwith, held the last service in Cornish in 1674. Cornish speakers remained on the farms. But the gentry spoke English; it was more 'refined'. Queen Elizabeth I could speak Cornish. In *Henry V* Shakespeare refers to the Cornish as a foreign-sounding lot. Some of the last people to speak Cornish were the elderly fishermen of Newlyn. Out on the sea, where no one could hear them, they used it for nautical terms. One popular story was that a fishwife in Mousehole, Dolly Pentreath, was the last Cornish speaker. When she died in 1777, the language died too. It was not passed down to the children. When that happened, the Cornish lost their separate identity.

The council house was like a palace after the caravan: all that space, stairs to walk up. Big bedrooms. Sally could not go into full decorating mode, as money was tight, so she was stuck with a lurid gold carpet. They had to go out of the window to reach the tiny garden, as none of the outside doors led to it. Martin wanted to be the only breadwinner, so Sally could bring the girls up properly. It made it difficult to keep his head above water, to feed the family, keep them warm and pay the mortgage. He needed bigger catches. He worked hard in winter, until the Atlantic storms stopped him fishing. Then he had to go on the dole and fill out the forms, with stupid questions about the size of your biggest catch – they didn't understand how different share fishing was to other businesses: the crew shared the profits of a good haul but on other days went home with nothing. There was never any money for Christmas; the girls never had big supersonic presents. One Christmas, Sally bought a dress in an Oxfam shop but prettied it up with different coloured sashes and things and put it in a huge, purple gift box. She recycled another dress. The sisters didn't know any better. They loved the dresses.

Martin had no salary, no wages. The skipper deducts an extra share for maintenance of his boat. If you can't find the shoals, word gets around and it's difficult to keep a crew. No one wants to work a fifteen-hour day in a winter gale for a skipper with empty nets. On mackerel runs Martin started going out on his boat with only one crew member. It was much harder work but they only had to divide the takings in half.

The trouble was, selling fish was becoming so fiercely competitive. Martin needed an ally with some business nous. A maverick like him.

2

Martin started to moor his boat in Newlyn so he could land his catch straight onto the busy fish market. He found Newlyn very different from the peaceful life down Cadgwith Cove. He battled through so much traffic to commute in. The port was jam-packed with a fleet of beamers, side-winders, longliners and netters, moored in a huge tier up to sixteen deep. Martin had to clamber over them all to get to his boat. The tier slid around violently in the wind when a south-east gale pounded the harbour wall. Crews slept on board the larger deck boats in bunks heated by a coal fire. On smaller boats they lived like rats in tiny, unheated cabins shared with the engine. They had to ride out the winter Atlantic groundswell with crested peaks as high as hills. Troughs were so deep that when you were at the bottom of one, the boat on the other side of a high wave would be invisible to you. You had to stare three waves ahead to brace yourself for an angry wall of water striking the boat.

Martin had to wait in line at the landing berths. He was landing to a fish merchant in Newlyn called Nick Howell, a real rebel. Nick was in a loud argument with the auctioneer at the fish market. He was always challenging them.

Martin first met Nick in 1979, when he walked into the Cadgwith Cove Inn. Plugger, his former skipper, was there nursing a pint of Atlantic, with his white whiskers and a frayed cap. Drinking with him was a long-haired, 28-year-old bloke in flares. Beside the burly fishermen he was lank and wiry like a wind-blasted moorland tree. Said he'd just been camping at the Isle of Wight music festival. His eyes were full of mischief. With his frowzy hair and snappy chat he was like a terrier. Nick. He was a lively talker. His fingers churned and kneaded the air. He spoke in a smoker's mockney, like a London cabbie, but was a public schoolboy from upcountry with a bit of attitude: a young man running away from the job his mum had found him in London, selling newspaper advertising. Nick hated the commute, so he travelled all the way to the other end of the line, near Land's End, where his only options were fishing or tourism. His van was outside; he was selling fish out the back of it. Apart from a few trips to Billingsgate fish market, he didn't have a background in fishing. Martin was excited to meet a new, young merchant, cutting his teeth. They were both in their 20s. If Plugger liked him he was alright.

'Who's going to handle your mackerel then?' Nick had asked Plugger.

'You reckon you can handle it?'

'I can handle every fucking one.'

'And pay us more than Stevensons?'

That was the real trick, of course. The Stevenson family were the Newlyn 'mafia'. They'd ruled the port like feudal lords since the 1850s. By the 1980s, Newlyn had become the busiest fishing port in England and Wales; next to the fine fishing grounds of the South-Western Approaches where the Atlantic storm surges were notorious. The Stevensons had the biggest fleet in the country with twenty vessels; they owned all the houses near the quay for their men, and farms on nearby land. They were the biggest catchers, claiming about half of all the fish landed in Newlyn; they also ran the auction, where their fish were sold first. They had the port in a stranglehold.

Martin and the other Cadgwith fishermen had to drive over to Newlyn market to sell their fish on their own - in four vans, each one half-full. Sometimes, when they landed late and couldn't get to market straight away, they had to store the fish in freezers in their sheds at home. They looked like amateurs compared to Stevo. Stevo was the nickname the fishermen used for the Stevensons. Nick offered to deliver the catch to Newlyn for them in his van, packed properly in slush ice, a mixture of crushed ice and seawater, like wet ice cream. The fish inhale it into their gut, their core temperature drops, and their shelf life increases to five or six days. He cocked his head to one side and grinned as he told them how his little van, packed with their fish, could drive straight past Stevo's auction, overtake their thundering container lorries, and deliver their fish first to buyers far, far away; even onto the ferry and across to the continent. The Cadgwith men, including Martin, agreed.

When Nick was at the fish market he darted in, dug down deep into the box, yanked out a fish, pored over it like it was a nugget of ore. He bobbed around. His brain was like a tuned engine. His hands were turning with his ideas: history of pilchards, the glory days of the purse-seine nets. He knew everything. He worked until the small hours and had big, entrepreneurial ideas. But the way the Stevensons dominated the auction angered him. They weren't interested in any new ideas he had. It suffocated him even going there. Martin sometimes went longlining during the small tides to catch ling and conger eels. When the conger season started in the Lizard, Nick told Martin: 'Imagine if we shipped it straight to Spain. God, they'd love the quality of conger caught here.'

He smiled to himself as he bypassed the Stevenson auction altogether and shipped all the Cadgwith conger to Spain. Soon Newlyn caught on too. Next season thirteen boats went with him as wholesaler. Word went around that there was a ballsy new fish merchant, a maverick taking on the Newlyn fishing dynasty. Martin saw how hard Nick worked, tearing around in his van all night, fighting his corner like a terrier. He knew he'd found a strong ally. He followed Nick to Newlyn.

In Cadgwith Cove withy pots were on sale to the tourists; wetsuited children flung themselves off the Todden into the sea. Martin found Newlyn very different. It was a forty-five minute drive to the bustling, brightly-lit port that stank of diesel, with shuddering trucks; it heaved with bodies, marching down the quay looking for work. Newlyn was on the up in the 1980s: beam trawling and deep-water netting were growing. The pubs were packed all night, and everyone had wads of cash. You'd do your twenty-three days at sea, then come back to party hard. You could buy anything you wanted to sniff or smoke. The Star Inn was notorious for it – the curtains there closed at the same time every night.

There was bottle walking, lifting barrels over the head. With the money coming in, lots of the local boys bought powerful motor-bikes. One lad, Ivan, went out all pissed up. His friends tried to stop him, tried to make him put on a helmet, but he rode straight through a car windscreen, out the other side and was killed.

The Swordfish in particular had a reputation as a hard pub, but Newlyn was no worse than other ports in this respect. People shook hands the day after a fight. It went to hell when the Grimmies invaded, locals say. They came off the ship in a brand new Burton's suit, then crawled back on board in tatters two days later, their wages lost in three-card brag. Burton dummies they called them, men whose options were to go to sea in a factory ship or go to clink.

The Stevensons' fleet was mainly beam trawlers, a hundred feet bow to stern, with a galley, sea shower, sea toilet. They went out for ten days and burned 10,000 litres of diesel a trip. Working these ships was nothing like the short trips on the smaller boats Martin skippered. There were dedicated sleeping quarters on the Stevenson boats: when you went to sleep you climbed through a hole into a coffin, with a little curtain and light switch. They had a three-hour trawl, during which they would drag the net for sixty miles. Slit fish bellies, scrape liver and guts away with the knife, wash off the slime and cover it with crushed ice. If your luck was out you would be constantly on deck, working from six in the morning to midnight. But the smells from the galley would be worth it. If there was a roast in you could smell it cooking down there when you were working on deck. The shitty taste in your mouth was salt, fish, diesel, oil, all that slimy, silvery glit on the bottom of the ocean, the sediment that stuck to the nets, you got covered in it. But through all that, with a galley on board, you'd be thinking, *I can smell that chicken.*

At sea there was no drinking, no smoking weed, none of that fucking lark; there's nowhere to hide on a boat and there's always something to be done. A few times men on the Newlyn ships claimed to find human skulls in the nets, kicked them out through the scuppers, the holes in the ship's sides. In the pubs they talked about one guy who pulled up his friend with the haul. Crabs had only left his spine, but he recognised his necklace.

These things made Newlyn men superstitious. You couldn't say 'rabbits' at sea, you called them 'underground race horses'. Its origin went back to ancient Cornish folklore where a group pursuing a rabbit saw it turn into a red-eyed witch. Couldn't take a pasty to sea. You couldn't whistle at sea, you whistled up a wind. If you saw a priest on the quay, you turned back. There were loads of things like that. The widows got a bag of fresh fish sometimes – they got lemons, monk, dabs. Fish straight off the boat was unbelievable, like eating steak.

All through the mackerel boom the Scottish skippers continued to flock to their Italian restaurant in Falmouth, La Cucina. It was by the Custom House Quay on Arwenack Street, just off the high street. A friend of Nick's from Cadgwith called Peter Tebbitt, who did the fish rounds, ran it; his girlfriend did the cooking. A French waitress, Marie Thérèse, was left to run the place. She was petite with dark eyes and gamine dark hair. She was from a fishing village in Brittany, so had grown up on quays, but she found the thick accents from the Scottish and Shetland fishermen hard to understand.

'Speak slowly,' she asked them. 'You're a foreigner like me.'

When La Cucina's owner married his girlfriend he held the reception in his restaurant. But he planned to drink so he needed someone to drive him from the reception back to Cadgwith. He tried to rope Nick in.

'You've got to meet this French girl,' he told Nick. 'She's fantastic.'

Nick drew up at the restaurant, checked himself in the rear mirror. His hair was like a clump of kelp, he looked knackered, but at least he had a tie. He walked in. The reception was mayhem. Thumping music. Everyone was completely pissed. Marie Thérèse reeled around in Tebbitt's dinner jacket, her hair sticking up like sedge grass. Nick watched her from the bar. When she went outside to catch her breath, he followed her.

'Hello, I'm the chauffeur for the happy couple.'

'Where is your Rolls-Royce?'

With a flourish, he gestured towards a flatbed fish truck with a three-piece suite for the newly-weds to sit on. She screwed up her face.

'So romantic.'

She forced him to decorate all round the side with cabbages. The bride and groom clambered on board. Nick let Marie Thérèse ride in the cab. Cars honked their horns. She wolf-whistled out the window. They laughed so hard; he couldn't keep his eyes on the road. Suddenly there was a hammering on the cab roof. The groom's puffy face loomed in the back window.

'Stop!' he hollered. He stumbled to the edge, in his wedding gear, and had a pee. The reception carried on at Peter's house in Cadgwith. Nick tried getting off with Marie Thérèse. Every time he leant towards her, she leapt to her feet to dance.

In the weeks that followed Marie Thérèse was won over. *Drink up!* She kicked the drunk, wilting skippers onto the cobbled street, washed the glasses, locked up. Then yawning, she followed the catseyes until the Penzance sign loomed out of the dark. He lived on the edge of Penzance, a short walk into Newlyn. In minutes she pressed her face to Nick's moustache. He tasted of coffee and tobacco. His eyes were seared with tiredness, but happy to see her. His flat was filthy. And tiny. It smelt of unwashed clothes. Plates piled up in the kitchen. He smoked a pipe! How old was he – 105? Still, her own place was a shoebox with students, who asked for French lessons. So she washed, steam-ironed, folded all his clothes and hung them up in the wardrobe, above his clutter.

'Is it weird being French in Cornwall?' he whispered to her.

'Don't you know anything?' she laughed, propping her head up on her elbow. 'I'm a *Breton*!'

Even though it was 2 a.m. and his catchers would ring him at 5 a.m., she told him about a festival where Breton fishermen trail blue sardine nets through the streets; the pretty local girls dance with white headdresses and full skirts. Men wrestle. Men on horses hold up flags. Many ships had been wrecked off the Île de Sein, the treacherous island near where she grew up. So Cornwall was like home, because she was a Celt. Like the Welsh, Scottish and Irish.

'You don't mind being in this smelly fishing port? With all these fishermen?'

She laughed again. 'My whole family is at sea. My life is the sea!'

Her dad was a merchant navy captain, her granddad was a skipper, her cousins were at sea and her brother-in-law was the Secretary of the Breton Fish Merchants Association. From the age of 9 Marie

Thérèse walked alone down a single track from her house that led to the sea. He noticed that she knew all about quays. He'd turn and find her talking to fishermen about weather and sea conditions. He lay awake chatting to her. The moon peeked in through the curtains. She crunched cigarettes into the ashtray. They never slept.

She grew up in Bénodet, where Brittany's prettiest river, the Odet, crossed by tiny, geranium-lined bridges, finally enters the sea. It is near Quimper, the cultural heart of Brittany, known for its cathedral and atmospheric old quarter with half-timbered fourteenth-century houses. She struck out to work the tourist season in London and soak up the 1970s music at the Isle of Wight festival, but ended up shivering on Oxford Street in a damp Father Christmas outfit.

After seven months living together, Marie Thérèse went back to Brittany as her father had Alzheimer's. Nick's cat tiptoed around as clothes piled up on the floor. The place became a tip again. It didn't look like Marie Thérèse was coming back. She had to look after her parents. So Nick phoned and asked her to marry him.

They were married in Bénodet. To pay for the trip Nick turned up with three tons of winkles in his lorry. He had been in high spirits on the ferry to Roscoff with Nutty Noah and other Cadgwith friends and employees. At Roscoff they all disappeared into the sunset, leaving Nick to load the bloody winkles himself into a Spanish lorry.

Marie Thérèse liked visiting Martin, Sally and the Cadgwith crowd. It was more like Brittany for her with the fishermen gathering at the Cove Inn and singing the old songs there. It reminded her of the annual, week-long Breton festival where

Quimper was filled with Celtic rhythm from bagpipes and girls in traditional costume. When the fishermen landed their catch, Nick would go down with all the cash and pay them in his van. They bought a converted place in Penzance, Vellanhoggan, at the back of the Barn Club. It was the early 1980s. Nick had just turned 30, was married, and had a house to pay for. Marie Thérèse helped him out. While she was pregnant she climbed the steep steps of his tiny fish loft off the fish market, down an alley beside the Star Inn. It was steeper than a ship's ladder and dangerous. She gripped the rungs. One hand clamped over her mouth, her cheeks pumping air. Jesus. Morning sickness. It was like the worst hangover she'd had, every bloody day. Nick's pipe was bad enough, but the stink of fish sent her bolting to the toilet. They bought from the fish market on Thursdays, then had long days packing them up; by Saturday polystyrene boxes were stacked high on the road, waiting for the Spanish lorry to turn up. There was no room in his loft. He needed to find somewhere bigger.

Everywhere he went in Newlyn the answer was the same: it was owned by the Stevensons. He even found derelict premises, but they owned those too. It felt as if the Stevensons had bought everything near the quay to make sure no one else could use it. Billy Stevenson's younger brother Tony looked after all the auctions, the property in Newlyn, all the stuff onshore.

Marie Thérèse had a plan to help. She invited Tony and his wife, Shirley, round for dinner. She would cook something nice, they'd share a few glasses of wine and it would be sorted. Then Nick came home at five and said he'd had a big fight with Tony on the market that day.

Billy only talked to Nick about pilchards because he was fascinated by Newlyn's history and felt nostalgic for the ancient pilchard fishery. There was still one working factory in Newlyn, called the Old Pilchard Works, owned by Shippams which salted and pressed them in wooden barrels. They had been selling to the same Italian family for a hundred years. Nick went to view the factory. It was in an awful state. None of the locals would touch it. Pilchards was a peasant product: head on, gut in, an ingredient in many pasta dishes, especially in Sicily. Little mobile vans rattled round the Italian mountains, delivering to the *alimentari*, backstreet grocers, in Naples, Turin and Milan, where tiered boxes of fruit and veg bulged all over the pavement.

In 1981 Nick bought the Old Pilchard Works. He kept it going. It needled Billy that Nick had bought a large, historic pilchards business right on the Stevensons' doorstep.

Nick started to hate the Stevensons. They were so powerful in Newlyn. He kept building up his Spanish customers; when they came over Marie Thérèse cooked for them. She learnt Spanish. She wanted to do something to help him. He was full of ideas and working hard. She just went along with it. Days went by just jiggling her baby boy. Red-cheeked, he gnawed on his chubby fingers. He woke up every half-hour in the night, teething. She pranced around the room in circles until dawn. Chywoone Hill was a sheer slope; dangerous for a pram. She cut a lonely figure heaving her baby up there in the mizzle.

Nick bought from a local fisherman, Freddie Turner. When you set your nets out in the ocean, you could come back and find that a trawler had gone straight through, or someone else had hauled your nets. Boats talked to each other, they sorted it out. Once,

Freddie saw a French trawler consistently ignore the markers and plough straight through his nets. He was livid. Freddie was into guns. He took his shotgun down to the quay, and when no one was looking he stowed it in the wheelhouse. Then he scoured the sea for the French boat. Soon enough it appeared and chugged straight through his gear. *Bastard!* This time he'd frighten the guy off, give him one hell of a warning. Freddie steered to within range. He stepped out of the wheelhouse and raised his shotgun, levelling it straight at the Frenchman. The Frenchman ducked into the wheelhouse. He then stepped out and raised a rifle with telescopic sights.

Freddie was pretty shaken up. He took the number of the boat. Marie Thérèse thought she'd write to the French skipper, to cool the dispute down. She called her brother-in-law, who worked for the Fish Merchants Association in Brittany, to find out the address of the French boat. But when her brother-in-law heard, he rang the skipper up. The guy nearly shat himself. Marie Thérèse's French contacts were proving useful.

After one argument with the Stevensons Nick was refused ice. Fish deteriorates quickly there on the hard, and without ice he would go bankrupt – and they weren't bluffing. Nick had seen how brutal the Stevensons could be. By controlling the ice plant, the Stevensons could ruin their rivals in one blow. What would happen to Marie-Therese and the baby if he went bankrupt? He lay awake staring at the ceiling. Martin's family and the Cadgwith fishermen relied on him to pay them on time. Without him they didn't eat either. He cursed Tony and Billy Stevenson, who had inherited money from their dad, when Nick had built something from scratch. At the pub, he'd boil over, shouting at people who mentioned their name – how could they treat people like this? He had to do something. So he fought back. Soon after that he decided to set up a new fish plant and buy his own ice machine.

Nick was driving up the Coombe, the valley leading out of Newlyn, when he spied a sign saying 'industrial units for rent'. It was the first industrial estate they had seen. The building was originally an eighteenth-century foundry. Someone had tried to revive it as a fishmeal plant in the 1970s and spent a million quid on it, but there had been a complete cock-up with the planning. The council had closed it down and now it lay derelict. Nick didn't think there was a hope in hell of getting in there as a fish business. And the planner had retired. Nick moved in and created a modern factory. He'd heard customers moan about tiny fish being hidden in the bottom of the box. Crews are supposed to weigh the fish at sea, ice them and box them ready to be sold at the auction. But the standard of box-fish stowage at sea varies with each crew member. Nick weighed and graded each fish by machine onshore so that wouldn't happen. To Marie Thérèse it was just another bloody idea, another bloody property he was buying. Ideas, ideas, ideas. Buy, buy, buy. It never stopped.

For the Stevensons it was like a declaration of war. He'd invaded their stronghold. He defied them by breaking away from Newlyn, moving up the Coombe and taking boats direct. Besides his mortgage, he now had to pay off a factory. He'd started with Plugger, but now had thirteen Cadgwith boats landing to him direct, and he brought it into the market. He had twenty employees in all. It was a lot of pressure. All white fish, only doing a tiny bit of crabs. He built the first lobster tanks in Newlyn. As the 1980s went on, Newlyn started to feel smaller and smaller. Everyone knew Nick would come up against the might of the Stevensons sooner or later. It happened in 1987.

As a fish merchant, Nick arrived at the quay auction early in the morning and bought fresh fish in boxes to sell on to outlets upcountry or in Europe. In 1987 £17 million of fish went through the Newlyn auction, £5 million more than its nearest rival auction at the port in Brixham in Devon, over a hundred miles away. The Stevensons were still the biggest catchers, the biggest buyers and they controlled the auction. They always sold their catch first, while it was still fresh. The smaller boats had to wait as their catches deteriorated in the warm air of a summer's day, the price dropping. Martin might joke about it, roll his eyes, but he wasn't the type to cause trouble. The Stevensons provided him with the cheapest ice and fuel, and settled his landing fees. Billy Stevenson was even a harbour commissioner. Martin could not afford to steam to the auction in Brixham.

Nick was more bullish. As he had expanded his sales to Spain and Belgium, he could no longer just sell the Cadgwith boats' catch. So for twelve years he had to buy fish from the Stevensons; he was frustrated that there were no other auctions or competition. How could the Stevensons be buyers *and* sellers? No other port in England had auctioneers and buyers who ran the ice and fuel too. Penzance Council, just over the bridge, set up a rival fish auction in 1985, but the Stevensons piled on the pressure and it closed down a year later and was let to the Stevensons instead.

Since the early 70s a fishermen's cooperative had been set up to improve the price of line-caught mackerel. It was affiliated with another cooperative which had other branches in Cornwall and

Devon's main port, Brixham. In 1987, when Brixham ran up debts, the Cornish fishermen broke away from Devon, investing in their own Cornish cooperative which tried to rival the Stevensons' auction at Newlyn. About twenty regular buyers began buying fish from the cooperative. In 1987, watching the growing problems with the Brixham cooperative, the Stevensons flexed their muscle again, and made moves to close it down. One skipper, Mr Rayment of the *Confide*, got into an argument with Tony Stevenson. Tony rang the market bell, gathered the buyers together and told them that anyone who bought Rayment's fish would be banned from the Stevensons' auction. For six months Rayment couldn't sell his fish at Newlyn market.

The Stevensons warned small buyers not to buy from the cooperative. The family would no longer supply them with fish if they did. The Stevensons accounted for eighty per cent of all fish sold at Newlyn so no buyer could ignore them. They also started demanding large 'cash bonds' from buyers as a safeguard against the buyer defaulting. These bonds would be waived if the buyers boycotted the cooperative and only bought from the Stevensons in the future. For one buyer the bond was five grand, for others it was thousands more. From the cooperative itself they demanded a bond of £30,000 to cover its bills in advance.

The message was clear – you take the Stevensons on at your peril. The family made their own rules, as they had done for over a century. Buyers drifted away from the cooperative at Newlyn until, in February 1988, it went bust and the receivers came in. The feud had echoes of the famous Cornish novel *Poldark*. Ross Poldark, a maverick, took on the wealthy, powerful Warleggan clan, who exploited the working poor. He formed a cooperative of tin miners to try to gain a better price for his tin. But the Warleggans were too powerful.

Nick helped Rayment out. He had to think on his feet. If the Stevensons were going to make life difficult for Rayment, he offered to travel all the way to Brixham, buy Rayment's fish there and ship it back to Newlyn to be transported out. Nick would buy fish from Scotland downwards. Ayr market was in the evening. He could buy fresh Megrim there and sell it for twice the price in Newlyn the following lunchtime. The big buyers in Newlyn all had good contacts in France sewn up, but Nick found the Spanish paid premium price for Megrim and Hake – so he sold them straight to Spain.

In May 1988 there was an Office of Fair Trading inquiry and the Stevensons' tactics were exposed. Their practice of demanding thousands in cash bonds from their rival, the fishermen's cooperative, disappeared. Suddenly there was oxygen in Newlyn again, and the market opened up.

Many years later the Stevensons were also revealed to be running a huge scam involving the illegal landing and selling of fish in Newlyn. The court heard that the skippers of their fleet had fiddled the books over six months. They had described high-value cod, hake and anglerfish, all with tightly controlled quotas, as lower-value ling, turbot and bass. The auctioneer was in their pocket; they falsified the records so no one knew at the time. Almost a quarter of the fish sold was illegal; the scam netted the family around £4 million. In the Newlyn pubs there was some sympathy, especially among those on the payroll. The Spanish and French have lied about their quotas all along, they said.

After the relatively peaceful Cadgwith years, Nick had discovered that Newlyn was a snakepit. The Swordfish Inn and the other pubs along Newlyn quay were full of fishermen who worked for the Stevensons. They rented the stone cottages that stretched back from the quay in a maze of tiny streets. You had to be careful who you stood next to at the bar.

One day Tony Stevenson bumped into Nick and said, 'You did one thing right.'

'What's that?'

'You married a Cornish girl.'

'She's Breton.'

'Exactly.'

Cornwall was always closer to the Bretons than the English. The ancient Cornish tongue was related to Breton. The Cornish practised distinct Breton-style sports: hurling with silver balls and bouts of wrestling. 'Kernow', the land of the 'West Welsh', was an independent country. West Cornwall was always different: Penwith and the Lizard were defiant outposts of Cornish Celts. English incomers in the sixteenth century were told bluntly 'I can speak no Saxonage', 'My, ny vynnav vy kewsel Sowsnek' in Cornish. There were no Saxon settlements with village greens, seen throughout south-east England. Only huddles of stone cottages with old Cornish names, just out of bowshot of their neighbour.

In Newlyn, the local fishermen remained wary of outsiders. Once a Frenchman came into the Swordfish Inn for a pint, and word got out that he had torn up a Newlyn man's nets. While he was drinking, a group stole out with spanners and hammers and smashed his wheelhouse and radar to bits. Just as they didn't like a Frenchman fishing their waters, nor were they keen on a man from St Ives Bay or – as Martin discovered – Cadgwith. They'd have a laugh and a drink with Martin but did not forget he was from across the bay.

The Swordfish was a good place to drink. One regular was local character Ben Gunn. He was a rotund, wide-eyed fisherman from Wick in Scotland, an experienced trawlerman, with a bass Scottish voice, thick as black treacle. Ben was thankful he was so short he could duck under a heavy twelve-metre beam and tickler chains swinging from a derrick in high seas. He'd carried a gun with him while fishing in the Icelandic Cod Wars.

He'd enjoyed Iceland but it was a struggle at times. Those days haunted him. The camaraderie was brilliant. You could get away with everything apart from murder. If you were up in front of the owners for doing something like fighting on the boat, then you'd be up in front of the managing director who'd be sat there with dark glasses because his wife had filled *him* in! It was run well but the money was crap.

They'd had Icelandic boats coming at them the whole time. The Icelanders would cut one of the warps, the two ropes that attach the trawl net to the trawler. The hawser would snap back onto the boat like a whiplash. It left only a single end; that was half a day's work gone. Jesus. More work, no fish, no money. The Icelanders rammed the trawlers, fired rounds at them. Ben Gunn didn't take any notice of the ice, just got on with what he was doing. He had a friend in Iceland at that time, Alfie Bothwell. They had been on a football boat together. They were caught in a hurricane off north-west Iceland in a place called Drangsnes, in the Denmark Straits. The weather was hellish. They couldn't get their breath. There was ice everywhere. Mountainous seas coming over the side. They had to get the big arm bobbins for the trawl on board by hand, get a rope around their belly and lash them down to the rail, otherwise they're in bigger shit then. The bobbins were solid rubber wheels or balls that held the bottom of the net open and helped it crawl along the sea floor. It all had to be lashed. The arse end was going under.

'Water coming! Get hold of something!' Alfie cried out.

Ben put his hands through the handrail and grabbed on tight. You got hold of any damn thing so long as it was iron. She dipped and carried him up. Green water flooded over the rail, washed everything off the deck. It wrenched his feet from under him, but he clung on, tightening his arms.

'If the boat's going we're all fucking going,' he cried out to Alfie.

When she came back up Alfie was gone. He got washed overboard. That was the worst moment of Ben's sea career. Christ. The feeling like nothing else. Jesus. After that he went on the drink for weeks.

But it wasn't the worst storm Ben was in. That was in the North Sea, fishing out of Lowestoft, with a man who had lost one ship with five hands. They heard on the radio that earlier that night a hotel rig had toppled over and sunk, with a lot of people lost. Then the storm hit them. Only three were allowed on watch: the skipper, the mate and Ben, who was the bosun. There was no autopilot. He had to shut his eyes on deck. They had great fish-holds on the deck. The sides were wooden pen-boards several feet high; they were bent like matches. He looked out, beyond the deck lights, into the churning waters. A dark wave was bearing in, a black wall racing straight towards them. Ben's body began to roll, automatically. The wave struck the boat midships. The water surged over the rail like a weir. It kicked his feet from under him, as the rubble of gear, nets and fish slid over to the other side. Ben held on. He knew exactly when. There's no amount of telling. It's experience. If you're not used to it, it's more dangerous. He could tell the deckhands until he got blue in the face, until they got a clout. Then they listened.

Off the starboard rail he saw a whale. Breaking the surface like a huge granite reef.

'I tore a nail off my big toe.'

He twisted back. Wild-eyed, the cook was stumbling onto deck.

'Is that all you're fucking worried about?' Ben snapped, turning back to sea. The great fish was gone. Had dived into the darkness of the sea.

He remembered fighting on the deck over a chord on a guitar. Ben put a crew member on his arse, but like an idiot he let him get up. Never told Ben he was a European boxer. Knocked seven bells of shit out of Ben. They were great pals after that. This was how things were. They sorted it all out. Always got on alright with most people. They hadn't had a drink for a month and a half when skipper decided to release some drink from the bond. You can't say any more than that. Just drink. They don't carry any bond now. They've never had it down in Newlyn. When the deep-water trawling went from Hull and Grimsby, in the mid-1970s, that was the end of the bond. In the bond was just rum, baccy and a few sweeties. They used jugs then. On a ship called the *Bell Gorm*. Pour the bottle into a jug and it went round, so you didn't spill it. Then there'd be someone who didn't drink and the others'd buy his rum up. So you'd have two noggins then. It never occurred to the boat owners. You had the crew of the ship, the mate of the ship came down first and he'd see who was sober or reasonably sober and he'd say: 'Yous, yous and yous are on watch. The rest of yous party time.'

Then you'd be on the deck the next day getting the gear ready for shoot. Maybe a net set, new trawl, new bobbins. It was work. Then you'd have a livener halfway through the afternoon, a couple of cans just to ease it off. Every couple of days you'd get a noggin, a small wooden cup, about a quarter-pint. If you got a big bag of fish you might get another one. Then after twelve days it was dry ship until you got in. Then more drink.

Ben was in Plymouth on a ship called the *Greechi*. The skipper was a mate of his. They were in the White Sea off the Russian coast. It was minus fifteen, a polar climate. The sea freezes from October. The ice floats. Like a desert of smashed glass. It can be five feet thick, then wispy contrails streak through it. He thought: 'I can't be doing with this shit-shovelling. That's not me. I'm not going to do time for no bastard.' So he left. He thought he'd hit Ireland. Then someone mentioned Newlyn.

'Your time for Newlyn is not yet,' his missus told him when Ben first wanted to come down in the mackerel boom.

It took him another eight years to get there. By the time he came to Newlyn in 1982 there weren't many mackerel left. The boom had gone. He never went mackereling. He'd never liked bloody mackerel.

The Swordfish was a great meeting place. He always made friends easily. Not many people Ben Gunn didn't know in the fishing industry. It took him ten minutes to get a job, but when he got down the quay he couldn't remember the name of the boat. He was standing at the end of the quay, thinking what the fuck was the name of this boat, when a white-headed bloke came up the ladder.

'Anyone down here looking for a mate?' said Ben.

'Yes,' the man said. 'Trewarveneth. Grimy Mike.'

So Ben went down aboard it and was mate there for four years. Trawler, side-winder, in the Stevensons' fleet. That first week he picked up £480 for five days.

It was brilliant. He used to have parties up in his fisherman's loft, just by the Smugglers' Inn, overlooking the quay. Mad parties. Music. Just drink. He's still got all the music. He'd always been into Buddy Holly, Roy Orbison, Dr Hook. The windows wide open, blaring out Creedence Clearwater Revival's 'I Heard it Through the Grapevine' across the harbour. It was open door. They still talk about it sometimes. There was a glass coffee table and a big suite. Drink everywhere. A little table on the side with more drink. And crisps if you wanted to eat. Even the taxi driver would be in. That was good craic. Joe Crow was there: an old-school Newlyn seafarer, full of yarns. Joe had sunken eyes and a shoulder-length, flyaway Cornish mane. He sailed a yacht to the Gambia and basked in the winter sun of Africa's smallest mainland country for a year, before returning to crew for the Curtises, a fishing dynasty from Polperro, on their triple-rigged boat which targeted quality fish like brill, lemon, dover and megrim. Joe was never without his tiny poodle, the Swordfish's official guard dog. He danced with Dee, who was living with a Dutchman called Tee Hee. You could tell she was destined to be Joe's. Tee Hee was named after his high-pitched laugh. He used to be on the *Fairplay* tug. Ben struck a deal with him: Ben gave him a basket of fish in exchange for two bottles of whisky for Ben's missus. It wasn't all

Newlyn, there were men from Grimsby and Lowestoft who came down here to work for Stevo. Ben knew them all because they were ex-deep-water men.

There dancing amongst them was Shaun Stevenson, Billy's nephew, son of Tony. He was young, ruggedly handsome and a bit wayward. He'd tried fishing but it wasn't for him, so he went out to Australia then got into building. He played the banjo and loved Country & Western music. He liked his drink and mingled easily amongst the ranks of fishermen in the pubs on the quay. He knew about boats, ships and seafaring. Never had no carry on.

When Ben first came down everyone in Newlyn had these parkas with fur. They had been trawled up in the nets from a wrecked container ship. Tales of Cornish wreckers went back centuries. The drunken landlord of the Jamaica Inn on Bodmin moor, Joss Merlyn, raved about wrecking, how they used lanterns round donkeys' necks, up on the cliffs, to trick ships into steering themselves onto the rocks. Then as sailors swam ashore, Merlyn and his band of wreckers smashed their faces with rocks until they drowned. Dead men tell no tales, he said. The clifftop lanterns and drowning of survivors was a load of shite, one for the tourists. But Cornwall was always a poor county, with a savage, stormy coast. In the Star Inn in the westerly town of St Just there is a sign on the wall that reads:

'St Just Prayer'

Dear Lord
We hope that there be
No shipwrecks

But if
there be let them be at
St Just for the benefit
of the inhabitants

Spoken in St Just Church by Parson
Amos Mason 1650

After the parkas they had the wood bonanza. A cargo of hardwood was washed off a container ship. Tons of the stuff turned up on the beaches and it took a pretty organised group of people to get it off the shore. People used the wood for winter jobs, building sheds and chalets.

In 1997, a freak once-a-century wave struck a container ship, and it rolled sixty degrees one way, forty degrees back. Sixty-three containers were lost over the edge, twenty miles off Land's End, including over 4 million pieces of Lego bound for the children of America. Ironically many of the pieces were nautically themed, so, after a heavy storm, children making sandcastles on the beach kept digging up miniature cutlasses, flippers, spearguns, seagrass and scuba gear. There were also dragons and daisies, but the rarest of all was an octopus. That Lego kept being washed ashore for years.

A load of Bic lighters washed off a boat once. Ben found them. But the best thing he found was hundreds of sheets, bound for North America, that were dragged up in the nets. They were divided amongst all the people in Newlyn. They were thick sheets, wrapped in plastic. They had to get all the damn sand out of them. He had them for donkey's years. During the 'stainless steel period', when another cargo was lost, Ben kept opening the cod ends to find condiment sets made in Korea. The crew got fed up with it in the end and kicked them out the scuppers. One trip, they were in deep water and they were picking up all these kiddie toys, radios, bags of knickers, bras and huge inflatable bananas. From a container ship. The crew blew one of the bananas up and strung it up between the aft gallows hanging it with panties, suspenders and bras. Billy went apeshite. He had no sense of humour. Ben never took it down, because he'd never put it up. He denied all knowledge. They were picking that stuff up for ages: even Christmas trees, big lions, tigers, thousands of radios and trannies. They dumped it all in the sea, back where it belonged, no good to nobody. They could have kept it. Sometimes seventeenth-century bottles turned up in the nets, or ammunition from the war. Ben kept six-inch shells under the stairs. There were loads of Second World War mines too.

Ben was used to Cornwall's fierce Atlantic storms. He was from the northernmost tip of Scotland, just by John O'Groats, where the Pentland Firth lies: Britain's most dangerous coastal area, with sixty-feet waves, fierce tides and currents. He remembered when a Siberian cargo ship, *Irene*, was adrift in howling winds up to a hundred miles an hour. The ship buffeted the Orkneys for days. The lifeboat launched from Longhope, an ancient Orkney settlement of thirty people where the Vikings sheltered. Snow flurries and rain beat down on them that night in the teeth of the storm. They searched and searched for the stricken ship not knowing that the crew was already safe; she had run aground and the crew were all ferried off on a breeches buoy. Still the lifeboat kept searching. Wick was the last radio contact. A hundred-feet wave capsized the lifeboat and it was lost with all hands. After they found the boat they burnt it; it was a Norse thing.

Trawling was Ben's first love. He'd always liked trawling. There's more to it than the rest of it. You've got to have a brain on you. There's not a lot of brainwork attached to beaming or even netters. If you got a trawl on deck with holes, snagged on a shipwreck or seabed rock, you've got to mend it or you're wasting your time. Netters don't do that. All their nets and gear are mended when they come ashore and new panels get put in. A trawl's got to be right to go in the water or you won't catch anything. You've got to have a little more knowledge. He'd cut the raw edges off the rope, melt it with a hot knife, ensure the strands were properly whipped. Then he'd carefully thread the twine through the hole in the net using a net-mending needle with an inner tongue. It's all finger and thumb, so you can't do it if you've lost those, like some of the guys in the Star who barely have ten fingers between them.

He enjoyed everything about trawling apart from fasteners, things on the bottom that catch the net like an old anchor. He liked that you go away for a few days and he could escape every-thing. He'd never been one for bills at all. Never bought anything on credit.

Ben was a fixture in the Swordfish. He could have a fight in there with someone and the next day he'd be on the piss with him. There was no malice or weapons. He could give someone a smack and that was the end of it. He'd never seen a place like that at all. Ben used to go into the three of them – Swordfish, Star and Dolphin. If you weren't happy there you could get in a taxi and go into town, raise hell in some pub, then leave. Have a grand tour. Down the Ship in Mousehole. Cheques screwed things up. It was mayhem when they were paid in wads of cash. Ben doesn't bother with Mousehole now, because people who bought second homes can't mix and they've buggered the whole village up. That's what he thinks anyway. But he's still got friends out there. Edwin Madron. Blewey.

Ben thinks superstitions are a load of shite. Doesn't believe a word of it. He put his faith in the man in the wheelhouse, the skipper. You just keep well away from the rocks. There's no need to be in there. If you are then someone is not doing their job right in the wheelhouse.

Once Ben should have been on a boat. But on the quay he bumped into a Grimmy guy, Vince, he used to drink with in the Star.

'I need to earn some money for my son to go to university,' Vince said. Ben let Vince take his berth and walked back up the quay. That boat never came back. She went down with all hands. *Margaretha Maria*. Hell of a time then. Vince's son was a little boy and it took him years to get over that. He never did go to university; stayed in Newlyn fishing.

Another time, long before, Ben was getting ready to set sail when his first wife Ellen tugged his sleeve. Her face was pale. She looked like a ghost. All the blood had drained out of it.

'Get your kit off there now,' she whispered. 'I've got a bad feeling.'

He took his bag off and left. The boat went down off Lundy Island. The crew used to drink in the Star. There was a ruckus about that boat and all. Money. Who owned it. Shit coming out the woodwork. It was a small beamer. The owners were not a

Newlyn family, but they were in the port a long while. Over the years Ben's memory has gone, but he looked up one night in the Star and saw her photo on the wall.

He doesn't believe in good-luck charms. Except one. When his wife died in 1997, he wore their front-door key around his neck, tied on a piece of string, which he wears to this day. The first mate on one of the biggest beamers had yanked him out the Swordfish, up the Coombe to a club known as the Belgian Congo, gave him his first spliff. He kept smoking at the wooden tables outside the Swordfish until his guttural drink-slaked growl faded away to a hoarse rasp. It was Jackie who saved him. She turned up in the Swordfish six weeks later abandoned by her ex, no good; she stayed with her mum in Stella Road. She'd been coming down since she was a kid. She was drawn to Ben, but could see he was in a bad way, drinking profusely.

'I can't take this drinking,' she said to him. So he gave up all the shorts and stuck to beer. They go at teatime now when the sun's out. Star is a fisherman's pub. In the morning you are ducking hooks. He can sit at the back when the boats are in and the Star fills up. Arguments over diesel prices, who's been out in the worst storms, who's got the port record. But he's not forgotten. He's an old fisherman. Someone calls out to him soon enough.

One of Martin's and Ben's friends was Stacey. He was known as 'Shouty' because his voice could be heard from a hundred yards away. Stacey was born at the end of Newlyn Green in 1966. At ten his parents moved to Pendeen, a remote village on the cliffs with a hidden smuggling cave, a '*vau*', that stretched far out beneath the sea. Stacey's dad, a stoker in the merchant navy, was a rampant alcoholic wife-beater who left when Stacey was 13.

Stacey didn't get on with his mum. At 14 he came home from school to find her with her throat and wrists cut. At first he just laughed, then he ran out in tears. His mother survived, but from that moment she died in his eyes. How could she do that to her three kids?

At 15 Stacey started coming to the Swordfish to buy weed and get pissed. It had always been a bit Bangladesh, full of funny characters with lots of cash. People on motorbikes, being whipped, being filled in with iron bars. You took your life in your hands. If any tourist came in, they'd think 'Fuck, I've made a mistake' and walk straight through. 'Straight-throughers', they called them. It was the only pub in Penwith where if a 999 call went out for a fight they sent nine policemen – everywhere else was three.

Stacey moved out the day he left school. It was 1982, and there were over 3 million unemployed. He and two mates found a twenty-five-feet abandoned caravan in a field on the cliffs: a square cigar tube with a bay window at one end. They jacked it up, put a hole in the floor, poured concrete in it to make it stable on the soft ground, then built a flue and put in an old wood-burner. They burnt anything. They built bunk beds, sawed a mattress in half to fit in and plumbed their toilet into an old tin mineshaft. They lived off the land, eating cabbages, cauliflowers and potatoes. One night six of them broke into a butcher's store in St Just and carried half a frozen cow along the cliffpath to Pendeen, getting covered in fat. They jointed it with a chainsaw.

It was a full-on lifestyle. Whatever money they got they spent in the pub, on weed. Stacey ended up on a downward spiral into violence, drugs and six weeks on remand in Exeter. There was no work anywhere except picking daffs and bulbs in all weathers in oilers.

'I'm looking for a berth,' Stacey told a mate in the Swordfish. He was 24. 'I don't want to go on a beamer, too much metal flying about.'

Stacey thought they'd be in every night, but they were out for a week on a gill netter. Working with the bland, mechanical smell of diesel all day made food mystical, overpowering, almost sexual. Garlic and onions, cooking a chilli, a ham sandwich at lunch. Night skies were insane. He was in the middle of a Christmas snow globe with shooting stars.

When a French trawler went through their gear, they threw a razor blade stuck into a potato and fired hand-held flares at its wheelhouse.

Most injuries happen when you are shooting or hauling the gear. One day the floats were flying out fast, pulled by the tide, ten feet apart. Stacey heard his mate yelp with pain, then saw him fly straight past, a blur of yellow oilskins. His mate's arm was caught in the net pound – dragging him straight for the corner of the bulkhead. In a second the rope would snap his neck. Stacey lunged forward and yanked him out. The two men slumped on the deck, wheezing. Stacey felt sick. Then he turned and smacked his mate in the mouth. From then on he knew damn well his crew mate would pull him out of anything. Even on terra firma he would have his back.

One rough night they were north-west of the Channel Islands in a sixty-feet groundswell. The boat went up and up, a wave like in a child's drawing, their bow tucked under. Stacey saw it crest above his head.

'There's no way we can haul gear in this,' skipper called down. They rode the next wave, down to the bottom of the trough. The fish house, where they stored the boxes, was twitching. Green water coming over the side. Then it was up their arse, which was the worst. Stacey leant against the wheelhouse wall, when it flipped over so far his feet lifted off the ground. The wall was now the floor. He said goodbye to everyone he knew.

He couldn't set foot back on that boat for a while after that.

He was out in another awful storm off Labadie Bank when an emergency call came over channel 16, the open channel. Stacey recognised the boat as Stuart's – an old friend from his Pendeen caravan days. Stuart was a larger-than-life Cornish boy, clever in his own way, who put his fists through walls. He'd tattooed Stacey; Stacey'd tattooed him. He heard the helicopter crew say: 'He's turning blue. He's gone.' Then the sea suddenly went calm, the wind stopped and the moon came out. He's still got the cutting from the paper. Stuart's death ended up like a ball of grit, a bolus that rolled around, eating him up.

Stacey had had enough. He was so broke he was sofa-surfing, with nowhere to live. Mousehole was a ghost town in winter, not a single light on along the quay. Fifteen of them decided to march on Mousehole and take over all the empty second homes. Some of them ended up squatting in a picturesque Mousehole fishing cottage, changed the locks. This was happening more and more. In Penryn, fishermen watched which boats never had their lights on and then homeless locals cut the boat rope, registered it for

£1 and lived on it. It could be a cruiser, yacht or barge – there's a whole community there.

Polruan is black at night, all 'tax loss' second homes, but a 65-year-old, Shian, lives in the woods nearby. Penzance locals watch the 'emmets', the tourists, to see when they go home from their holiday. A group of Kernow activists set up a website telling squatters how to check at the Land Registry for £3 if the owners lived a long way upcountry. Cornwall has 14,000 second homes. The locals are all on minimum wage. One of the poorest regions in Western Europe. It even gets bailouts from Poland.

Stacey started working as a builder. He refitted the kitchen in one Mousehole cottage four times in ten years, installing the old one at his own place each time. Some pay him on time, some are so rich they don't see the urgency. Stacey worked with a very good carpenter in Mousehole called Andrew Blewett, or Blewey.

Stacey's kids can't afford anywhere round Newlyn. Their friends squat in caravans. One was found living in the caves on the cliffs of Tintagel. Tintagel, home to King Arthur and the myth of the knights of the round table.

3

These were Martin's friends in the Swordfish in Newlyn – Ben Gunn and Stacey.

They'd both given up fishing: they loved to hear how Martin kept experimenting with ring nets. He was in his element, able to go whenever he wanted to, because a ring net worked in the dark or between two lights. The first challenge was to find a way of doing it with one boat. (Usually a pair of boats would encircle the shoal and come together with the net ends.) Martin planned to throw a weight out, a drogue, a sea anchor, which would drag the net into the sea. The bottom of the net would sink down attached to weights, while the top of the net stayed visible on the surface, held up by a string of tiny yellow floats. Keeping an eye on the floats he would then steer the boat in a full circle, coming back to pick up the net end. Once the shoal was surrounded he would swiftly heave in a rope threaded through the weighted edge on the seabed. This would close the net from below, catching the fish.

So much for the theory. When he tried it, strong winds kept blowing his boat into the encircled shoal, so the net became tangled up. 'You might as well hit me with a plank of wood,' he told Sally. 'How the hell do I stop that happening?'

He tried putting a propeller over the side, to thrust the boat back against the wind, keep it out of the circle. To his amazement, it worked first time.

But then there was more bad luck. One day his crew mate, Jamie Fletcher, got his arm stuck in the hauler that wound in the net. Jamie's arm broke; he couldn't fish for weeks. Few fishermen take out insurance because the brokers won't touch them. The job's too dangerous. Martin found it hard to replace him; he didn't know anyone else who had any experience of ring-netting.

He was still optimistic, but he needed a change of plan. Ring-netting was too labour-intensive; he couldn't do it without crew. So he had to improve the size of the catch.

Martin was landing his crabs at Harvey's in Newlyn, a third-generation Cornish family who had the crab market sewn up. In their warehouse on the harbour they had live tanks which could take thirty tons of shellfish. He bumped into other young crabbers there.

'Where are you from?'

'Cadgwith. You?'

'Porthleven.'

Porthleven was a bloody difficult harbour to land in, wide open to the south-westerly storms, with a hazardous ridge of rocks under the water called the Mopus Ledge or the Iron Gates. Like a Kraken of black granite the Mopus reared up out of the green-grey water. The gales and tide dragged boats onto it. For many sailors, the white spray breaking over those rocks was the last thing they ever saw. Newlyn was sheltered from the south-westerly gales so Porthleven men landed there in winter.

There was one lad at Harvey's called Edwin Madron. He was Martin's age and came from an old fishing family in Mousehole. He knew Ben Gunn too.

'I worked a ring net with my father in the sixties,' Edwin said, when he heard Martin was experimenting with them. He took Martin down to his father's fish loft in Mousehole. Jimmy Madron was a terrific rogue, famous and infamous and great fun down the pub. No one was an angel, but he was a devil. He had six children from two wives. Martin listened to his yarns about how he used the ring net to catch herring over in Ireland. Jimmy was calm, with lots of charisma and warmth. Shoulder to shoulder with him, in the pub, his voice rang out to the beams, huge and tireless. Jimmy could tip the pints down. His brother Jo was always singing and whistling. At the turn of the century five of the Madron family were lost coming home from the North Sea. His grandfather lost five nephews off St Ives before he himself was lost in Plymouth. Once Jimmy, Jo and their brother Edwin were pilchard driving off Plymouth with their father. Jo fell over the side and brother Edwin went after him. He couldn't swim and got into trouble. Jimmy jumped in to help. The crew had to hold the wheelhouse door shut to stop their father diving in after them. Jo and Jimmy survived. Edwin drowned. Years later, Jimmy would name his son after his dead brother. The Madrons knew their fair share of tragedy.

Martin got to know Jimmy and Jo Madron, and kept picking their brains on how they had worked the ring net.

The net was an absolute monster. It took three men just to lift it. It was 160 fathoms long and twenty fathoms deep. The

Madrons' boat, the *Renovelle*, was only a forty-footer. The bigger ones were more than fifty feet. Stevo's trawlers were over seventy feet. When the Madrons towed the *Renovelle* into Mousehole she had no numbers, no wheelhouse, nothing. So they took her up to Canners' Slip and built a wheelhouse on her and painted the numbers on. Jo kept it immaculate. The engine room was polished clean. Around the external exhaust exits you get carbon build-up: Jo had even polished the carbon so it was shining like a wet pebble. But the Madrons seemed rough and ready compared to another Mousehole boat, the *Spaven Mor*, whose crew wore jackets and trilbys to go to sea.

'There are smaller boats with bigger nets,' Jimmy explained. Martin wondered if he could cut the net down. Jimmy helped Martin get his head around whether that would work.

'When you have surrounded the shoal with the ring net everything is swimming in it,' Jimmy said. 'It's not like a cod end when you are lifting it. You bring it in close until you've got the shoal up so tight you can scoop them out. Be careful. Let them swim. If you keep pulling and pulling too much the shoal will decide to die or dive, and their sixty-ton weight will pull your boat over.'

When Jimmy and Jo started off with their home-made ring net in Ireland, they chased the herring with tiny eighteen-feet toshers which filled up. They took 500 stone but there'd still be 700 stone in the net. When they opened the net up the herring just sat there. Herring are dozy. They wouldn't scoot out. They had to try to cuss them out of the net, the dozy, dopey little things! Herring are thick, Martin thought. Martin was not having much luck finding grey mullet so he asked if they caught any pilchards. They said they'd tried but they didn't catch many. There may not have been that many around then. Maybe there'd be more now.

One day when Edwin Madron was 14½ his father turned up at his school.

'Could you send the boy home? We're going on holiday,' Jimmy asked the headmaster.

The headmaster gave Edwin half a crown. 'You've got to go home,' he said. 'Your father's booked a holiday.'

Edwin was excited. He took the bus home, walked up the garden path with his coin. His stepmother handed him a bag and boots.

'You're going to sea.'

That's when he started. He died for six weeks. Seasick. God.

'Get up, come on, you've got work to do,' Jimmy yelled and gave him a tap with his boot. You still had to do your work, didn't matter how bad it was. Edwin saw a different side to his old man. He thought he was a bit of a grumpy old bugger really. He kept longlining for skate and ling or sometimes ray. He worked all week and then Monday morning started again. The *Renovelle* became a family boat for Edwin and his three brothers, Stephen, Shaun and Baden. They slept and worked together on it. All the crews would stick with a boat for seasons. You stayed over the pilchard season, when you finished that you'd go longlining and stay for that season. From 15 Edwin was drinking in the Ship in Mousehole with his father and uncle.

'Are you sure you are 18?' the landlady used to say.

'Yes. Ask father.'

'Yes, he's 18,' Jimmy nodded.

Then on his 18th birthday, they went in the Ship and never gave it a thought.

'Give the boy the best of the house for his 18th birthday,' Jimmy said.

'Get out!' the landlady yelled. 'You've been telling me lies for three years.'

So Edwin was barred for his birthday, the day he went from a boy to a man.

He was making £200 a week, good money for a young man. After he was paid in cash he went down the Ship, up the Legion, to the Coastguard. They'd go around three times in a taxi, then down to Newlyn to the Star and Swordfish. Everyone had their little favourite spots. If the craic was good you'd stay there.

In the mackerel boom, you couldn't get 1,000–2,000 men working in one bay without them falling out. They were wild days but good days. In the old days you went outside and had a fight and sorted it out. It was like the Westerns. Edwin's cousin, Jo's son Jonathan, was called 'Guns', because wherever they went if they downed beer, they ended up fighting. They fought at weekends but were best pals all week. Penzance against Mousehole. Edwin gave up fighting after a while because he kept losing.

'Take it outside,' the landlord said.

'Give him one,' two or three would say.

It was mainly like two children pushing each other around until someone said: 'That's enough of that, let's go and have a pint.'

No one was arrested or put in prison for the night or sent to court. What a waste of bloody time. The community policed itself. Edwin had known the local bobby, Johnny Green, from when he caught him stealing apples as a kid, grabbed him by the scruff, clipped him round the ear and sent him off. Everyone in Mousehole and Newlyn loved Johnny Green. The police were from old local families and knew when to stay away.

It was a close-knit community and at one time they were all involved in smuggling. Mousehole men manned the boats that shipped in contraband from France and the Channel Islands. They hid casks in their cellars, kept watch or lent a horse to carry it inland. They were nicknamed the cut-throats, after Martha Blewett, an elderly local who in 1792 walked up the hill from Mousehole to Paul churchyard and was robbed, her throat slit from ear to ear. A 26-year-old fisherman was tried and hanged. Mousehole remained close-knit, with secrets covered up. Some older men liked to play the Cornishman, told the visitors stories so they'd buy them a drink.

At Christmas the children formed a procession of lanterns held aloft, some shaped like fish, and paraded through the cobbled streets, all wrapped up in woolly hats and scarves, clapping their mitts and singing songs to the sound of drums and bagpipes. The harbour was lit up with strings of coloured lights, reflected in the water. They roared out shanties in the Ship, no one louder than Jimmy and Jo, who raised a glass to Old Tom Bawcock, a seventeenth-century local fisherman. The legend tells how the residents of Mousehole were starving after terrible winter storms had trapped them in the harbour. Tom braved the stormy sea and caught seven sorts of fish, enough to feed the village. Everyone feasted on stargazy pie, baked with herring heads and tails poking out as if they were diving through the pastry crust. The lanterns were floated out to sea.

Harbour Lights started in 1963 with one string on the pier. All the tourists flocked down at Christmas until as many as 30,000 came through. Thirty people alone were needed for the lights. A committee was set up and infighting took place between the busy-body incomers and stubborn locals. Letters as long as a football pitch spelt out 'Merry Christmas', hammered down with a hundred stay ropes against the winds. A Celtic cross out on the rock, St Clement's Isle, was tricky to repair, as the harbour was sealed off with wooden baulks for three months of winter; the only way to fix the cross was to paddle out there. At Christmas, to the village, Jimmy was the life of the party.

But that's how he was with everyone else. They never saw the other side of him. Sometimes he was a grumpy old basket with Edwin.

'What the fuck's it doing there?' he'd shout, shoving gear around on the boat. That was his way. They never took no notice of him. Jimmy had started off as a boy out with his father and was left the family boat when he died. Jimmy came home sometimes half-cut and wrestled with his kids as fathers do. As he got older, he was a grumpy old git. His kids nicknamed him Barney because of his temper.

Jimmy took Edwin far out to sea. They went thirty miles out in a dangerous stretch of water off the Bishop lighthouse, north-west of the Ship. So dangerous that in one savage storm in 1874 the waves were a hundred feet high and reflected the light back into the terrified lighthouse keeper's eyes. But Jimmy wanted to go out further as that was where the fish were. His boat only had two Kelvin engines, fuel drinkers with small tanks – fifty miles was their limit. Years later when the fishing got better, they put in a new Kelvin with a blower on it, for 150 horsepower. With bigger tanks, they could go further. They started travelling a hundred miles out for longlining. They were catching pilchards, mackerel with the big net.

That far out he had to take care of his boy. His family was on board that boat. He was a clever man at sea though, and made sure his boy was never lost. It was before they had all these implements, the plotters and charts. Jimmy made his boy learn where all the hidden reefs were. There was a ledge in Lamorna called Lee House. It's a bad one if you don't know where it is. It's like the Runnel Stone only smaller. Sometimes as they drifted into Porthleven, Edwin looked down to see the Iron Gates, the fisherman's name for the dreaded Mopus Ledge. When the sea's running in and there is a great swell, people talk about how high the waves are but when the troughs are deep they uncover the hidden reefs and that is equally terrifying. The Iron Gates reared up in front of them as high as a house, a wall of black granite that broke the surface and kept coming. It was scary. You wouldn't get away from it if you were in there amongst that. It'd be a lovely fine day then all of a sudden the sea came from nowhere and broke over it, seethed over it; it couldn't go anywhere so it just boomed and shot up. Boooff!

All the reefs are dangerous if you didn't know where you are. Wolf Rock is particularly treacherous as there is deep water all around it – so they built a lighthouse on it. The Longships is a cluster of rocky islets, one and a quarter miles west of Land's End. Inside the Longships there was a small, flattish rock called Kettle's Bottom. If you didn't know it was there and went over it you'd hit it. There's another one they call the Shark's Fin.

When you get a big tide almost all of these black granite rocks sink beneath the surface. Strong currents and tides drag boats onto them: the jagged edges can tear a hole in a steel hull. Below them on the ocean floor lie wrecks of all ages: minesweeper, schooner, U-boat, steamship and cargo ship. It is one of the most dangerous stretches of coastline in Britain.

Jimmy took Edwin out one day a mile from Gwennap Head, near Land's End. 'I'll show you the Runnel Stone reef,' he said. The Runnel Stone is a dangerous reef that has sunk many vessels. Years ago the top used to stick out the water, then a coaster hit it and knocked it off so you couldn't see where the reef was any more.

It was a fine day when Edwin went out there with his father; the surface was calm. So they came close to it, when the tide was slack the water still built into a breaking wave and ran down past it. It was a hairy feeling, all the seaweed was waving its fingers at him, like something was calling him. He felt his stomach churn.

The Runnel Stone buoy marked the position of the reef with a bell which clanged with the motion of the waves and a whistle set in a tube let out a moaning sound when the groundswell was high. Ding ding, woo-ooh. Woo-ooh. It was an eerie groan which could be heard clearly from the boat as it drifted in from the sea towards Gwennap Head.

'There, look up,' Jimmy said.

On the high ground above Tol-Pedn-Penwith, the holed headland, were two twelve-feet cones, one red, one black and white. You had to always keep sight of the black and white one. If it was completely obscured by the red one, then the boat would be directly above the Runnel Stone. This is what Jimmy's father taught him. Now he was passing it on. As the boat drifted nearer to land, Edwin studied the face of the cliff. At first it looked like any other cliff. A cave led in from the sea. A metalliferous vein streaked the opening from east to west. Water boomed and hissed inside. The roof had fallen in leading straight up a hundred feet to the cliff path above. This was called the Devil's Funnel and the opening was about seven feet from the verge of the cliff. A pile of large cubed blocks of granite stacked up to form the Chair Ladder, where climbers perched.

Jimmy was a good man on course, distances, tides, because that was how he was brought up. They'd be a hundred miles off, Jimmy would tune the radio in until he heard the series of bleeps. Two longs and a short.

'That's the Wolf,' he said.

The signal was coming from the Wolf Rock lighthouse. Every lighthouse has a different signal. Then he turned another dial until they could hardly hear it. When he had found the signal's weakest point he could determine what direction it came from. On the chart he plotted a line from the Wolf Rock lighthouse, which is eight miles off Land's End on the way to the Scillies. He then did the same thing with the signal from Newlyn lighthouse, on the south wall of Newlyn pier. Again turning the needle until the signal was down to hardly nothing, he used the two points to mark their position on the charts. Jimmy could come in using any lighthouse signal: Round Island, the Bishop or Land's End. That's how they did it in the old days. The stars all looked the same; they weren't going to tell you where the reefs were.

'It's your watch,' Jimmy would say.

There were always two on a watch. One watching the engine to make sure nothing was leaking. If there was water he pumped her out. The other would steer. Then they would swop.

'East by north half north.'

Edwin steered by that and Jimmy would clunk down the ladder to his bed. There were four or five watches. Then it'd be the last watch.

'Right,' Jimmy would say. 'Last watch. Call me at six in the morning.'

Edwin would call at first light and Jimmy would get up. Everyone else would follow. Out on deck it was thick, thick fog. They couldn't see beyond the bowsprit.

'Stop the engine,' Jimmy called.

They all shuffled out on deck. The white mist was a curtain all round the boat. They couldn't see where the sky ended and the sea began. Beyond them could be a whole bustling port, an open sea or a wall of black cliff.

'Now everyone listen.'

They couldn't see each other clearly it was so thick. Then all of a sudden they heard it. The eerie groan floating across the water. Then the ding ding of a bell. It was the Runnel Stone buoy, with its bell and groaning whistle that rose and sank with the waves. Ding ding, woo-ooh.

His father was a very clever man, Edwin thought. They had been steaming in for sixteen hours and couldn't see a thing. But he'd brought them to the tip of Land's End. They were all like that though.

Fishermen were very superstitious. They had all lost someone to the old grey widow-maker, the sea. One powerful old tradition about the white mist, kept alive in Sennen Cove, was that it was a spirit of the sea. A thick fog bank formed over Cowloe Rocks and slowly spread over the bay like a barrier. It emitted loud whooping sounds, so locals gave it the name the Hooper. At night it glowed with rising and falling showers of sparks. This fog spirit was a guardian that warned of approaching storms.

Another strong local Cornish legend had it that the Runnel Stone and the Seven Stones reefs were the tips of hills from the lost land of Lyonesse which lies beneath the twenty-eight miles of treacherous sea between the Scillies and the mainland. Beneath the Seven Stones lies its capital, the City of Lions; fishermen have trawled up evidence of old buildings there. The land had 140 churches. There are streaks of black in Mount's Bay that you can see at very, very low tide. They are stumps of petrified wood. At one time St Michael's Mount rose from a low-lying forest of oak, hazel and elder which stretched from Cudden Point, a prominent headland, to Mousehole and was covered by sea at the end of the Neolithic era. Gwavas Lake, off Newlyn Harbour, was a lake within the forest. All this was part of Lyonesse, which legend has it was drowned in one single cataclysmic night. There was one survivor, a Trevelyan, whose white horse carried him to Perranuthnoe on Mount's Bay. The Trevelyan coat of arms has a white horse emerging from the sea. The Cornish name for St Michael's Mount is Karrek Loos yn Koos (Grey Rock in the Woods). The pine trees around it today are only 200 years old. Since the twelfth century the Cornish have clung to the myth that King Arthur lies sleeping in Lyonesse, below the reef.

The Welsh have Owain Glyndwr, a hero awaiting the call to return. The Cornish have Arthur. So when a Welshman said he was a descendant of King Arthur and he sought to seize the English crown, the Cornish backed him: Henry Tudor. His advisers were greedy; once crowned, he taxed the Cornish and suspended their government. Fury erupted amongst penniless labourers. From the Lizard came a William Wallace figure: a blacksmith from the village of St Keverne, Michael An Gof, who stoked their rage until 15,000 men marched on London. This army surged through the West Country all the way to Blackheath, just outside London. From their camp they could look down and see Greenwich Palace. Their bodies were hardened from heavy work down the mines. They pulled the longbow right back and picked off the king's men. They fought with pitchforks, scythes and staves. It took an army of 25,000 to defeat them. An Gof was hanged, drawn and quartered.

One night the Madrons were involved in a mysterious incident in Mousehole. There was a painter who lived there called Justin Blake. (Blake was not his real name. He'd changed it because his real name was shared by another more established painter.) He just turned up in Mousehole in the 1960s. He lived next door to Edwin's friend Butts, a fisherman, in one of the narrow cobbled streets of tiny fishermen's cottages near the quay. He used to chat to Butts most mornings about painting. He was an arrogant, bombastic character with a very short fuse. Butts noticed that he drooled out the corner of his mouth. He was a nut. Complete bloody loo loo. He definitely had mental problems. But his paintings were brilliant. He was always down the slip in Mousehole painting. The first thing he would do was write the price on the back of the canvas. Twelve quid. His wife was so lovely, they wondered how she'd hitched up with him. He had a couple of lovely kids.

One night Blake was in the Ship Inn. Edwin was there too with his brother Stephen. Blake would lose his temper just like that. He rowed with everybody. He was like it all day long, all the time. That night the painter mentioned something about Stephen's wife and he took it the wrong way. An argument broke out. Blake threw a bottle. It turned into a nasty fight. Edwin tried to wrestle Stephen off Blake. The Ship Inn was a dangerous place to pick a fight with a Madron; the fishermen closed ranks. Behind the Ship is a labyrinth of dark lanes and granite fishermen's cottages. A barricade was put up to stop the police coming down one of the cobbled lanes; a small band of men pursued Blake back to his house. Butts was with four others over at Jo Madron's house and they heard screaming and shouting. Then a loud crash. They thought the kids had gone through the window. In fact Blake had seen the mob coming up the street after him, battering down the door, and he had thrown the television out of the window, down on top of them. Then a big flowerpot with a sapling in it. He hurled out glass jars with brushes in them. They shattered on the cobbles. Still the mob kept trying to force their way in. Butts went over and joined in. They dragged Blake down the stairs, duffed him up, gave him a slap. Then they marched him down to Penzance station and put him on the train to anywhere.

'Don't come back to Mousehole,' they told him.

Justin Blake went next to St Ives. Some say he went on to St Just, the old tin-mining town out on the cliffs. Then suddenly he disappeared. His wife assumed he had just pissed off. He was missing for fifteen years. The mines in St Just went out under the sea, so far that the miners could hear the shingle and waves roaring above their heads. When the mining ended they started capping out the mines. There were lots of bats in those mines on the cliffs. There used to be thousands of them in Mousehole harbour in the summer evenings. A whole swarm of them. The hiss of beating wings. A skein of black smoke fell down out of the gloaming. It was like a horror film. Then one summer there was silence. The sky was clear. They had capped the mines and the buggers died inside. The caps were stainless steel made up in Camborne. They were supposed to be made with slots so the bats could get out. Bat experts started going down on a rope like potholers, to check if there were any down there. Butts knew two or three of them. At the bottom of the mineshafts there was nothing but bones from dead animals – rabbits, foxes, badgers. Then they went down one particular mine.

'Normal, full of bloody bones,' one man called up. 'Hang on, there's a skull down here. And a skeleton.'

That's when the police were brought in. It took them twelve months to find out who it was. It was Justin Blake. It's more than likely that he pissed someone off. Maybe they killed him and chucked him down the mine. Or chucked him down the bloody mine while he was alive. The police never found any evidence to confirm he was killed; but after fifteen years there wouldn't be any evidence left. Just the skeleton.

Another Mousehole character was Trevelyan Richards, known as Charlie or 'Whackers'. He used to go down Newlyn quay, pick a rat up and stow it in his pocket then release it in the Bath Arms in Penzance where all the fishermen were. One night he came across a snoring skipper and blew a wad of snuff up his nose with a straw. He was always joking around. Whackers was a very big man. He came aboard to captain the *Carradick* when another skipper took six weeks off due to back problems. One trip didn't go well. He steamed down for skate deep off the Bishop; then turbot, a well-paid fish, off St Ives; then finally a bit deeper into the Irish Sea the crew had baited hooks up ready to shoot in the dark and all the electrics failed. They said they'd better go home. Whackers leant out the wheelhouse and muttered swearing which was Shakespearean in its depth; he cursed beautifully. Later, sat in the galley, he picked up two curved horn-handled knives up to his temples and said: 'I'm cursed by the Devil. I've never been lucky.' Then he said: 'You know, I'll never make old bones.'

Whackers was a key figure in one of Cornwall's most powerful sea stories. One December afternoon in 1981, Leon Pezzack, a telephone engineer from Mousehole, put up the Christmas lights around Mousehole quay. By evening there was a hell of a storm. The wind screamed. Waves battered the quay. It was a force twelve on the Beaufort scale. Rumours spread that the Penlee lifeboat, *Solomon Browne*, was wanted.

A cargo ship, *Union Star*, a coaster, had suffered engine failure eight miles off Wolf Rock. It was being blown across Mount's Bay towards the rocks of Lamorna. The rockets went up. The crew of eight lifeboatmen launched from Mousehole determined to rescue the crew of the stricken coaster. They were local volunteers, mostly fishermen. The coxswain that night was Stephen Madron, brother of Edwin and younger son of Jimmy. Leon's childhood friend and best man, John Blewett, was one of the crew. 'Whackers' Trevelyan was the skipper. Nigel Brockman had volunteered with his young son Neil, but Whackers told Neil to stay behind; the storm was too fierce to risk taking a father and son on one boat. It was a black night with sixty-feet waves and hundred-mile-an-hour winds. The *Union Star* was pitching violently and drifting towards the rocky coastline. The Sea King rescue helicopter was sent from Culdrose air base, manned by an American pilot. He radioed to ask how many people the skipper wanted taken off.

'Only one woman and two children.'

The skipper of the *Union Star* had picked up his pregnant wife and two teenage stepdaughters so they could be together for Christmas. The Sea King pilot flew in so close. The helicopter hatch door was prised open. The wind screamed like a buzz saw. The winchman, in his harness, pitched out into the teeth of the storm. He swung in wide arcs, blown back and forth like a kite. Below him the coaster listed, then fell away from him into deep nine-metre craters. Mountainous grey waves broke over the hatches and the bridge. A crewman staggered on deck with a female. He pinned her against a locker, to stop her being swept overboard. The winchman could see her pink deck shoes below him. The fifty foot aerial mast lunged up at him like a huge rapier. It whipped across, narrowly missing the rotorblades. But the chopper kept being blown backwards in the dark. The ship's anchor had snapped and now it was drawn towards the rocks. The helicopter was being blown into the cliffs. It was too dangerous. The pilot gave up.

Leon was on the cliffs watching. He led coastguard volunteers down to a zawn, a big cleft in the rocks, which dropped straight down over a hundred feet. He could barely see because the spray and spume were being blown into his eyes. It stung with salt.

Leon was lowered down the zawn on a rope for a better look at the rescue scene below. He scrunched his lids up tight; buried his head in his chin. The noise was terrible. He couldn't hear himself speak. Shingle was thrown in his eyes.

It was all up to Whackers and his crew from the *Solomon Browne*. When they arrived the *Union Star* was well into the shallows. It was in a V-shaped gully. There were only minutes left. The lifeboat crew threw ropes over and tried to come alongside. Seventy-feet waves crashed against the wheelhouse. The huge cargo ship was shaking. Then a massive wave came out of nowhere. Trevelyan turned astern. The wave picked the *Solomon Browne* up and dropped her on the *Union Star*'s deck for a moment, then lifted it clear again. She was back alongside immediately. Incredible seamanship. Four people in orange life jackets jumped across into the open arms of the lifeboat crew. They sent a signal back that they had got four off, including a woman, and were going back for more. Then the radio went silent.

Leon saw the coaster turn upside down against the cliffs.

Back on the clifftop a crowd had formed. A rumour went round that the lifeboat lights had been seen steaming towards Newlyn. The fishermen, including Billy Stevenson, ran down to Newlyn quay. She never appeared. Newlyn skippers, checking their moorings, called out on the radio waves to Trevelyan. The lifeboat radio stayed silent for hours.

Three lifeboats set out to search for the *Solomon Browne*. The Sennen lifeboat found itself climbing a vertical wall of water, as steep as a churchtower. The Scillies lifeboat surfed for quarter of a mile on a single wave. The Lizard boat bucked through such deep troughs its men were thrown mid-air. The shattered crew returned to find their port keel had split down the middle; they were half full of water. Their rail had been beaten flat by the waves.

Wreckage of the *Solomon Browne* was sighted near Tater Du: broken timber, air-boxes, a blanket. It was unmistakable. Rumours of the debris went round Mousehole. The families were now dreading the worst. Shocked and shaken they went from house to house, visiting each other, crying, making cups of tea. The streets of Mousehole were full of young and old, all in tears. The men were characters in the village and the loss of one would have been keenly felt. To lose all of them was too painful to bear. They sat in silence together. Twelve children had lost their fathers just before Christmas.

Trevelyan's body was found floating between Tater Du and Lamorna. Then Charlie Greenhaugh was found in Lamorna Cove. When Jimmy Madron pulled the wreckage of the lifeboat onto the slip all he found was his son Stephen's Breton hat. For years he'd taught his sons how to avoid the reefs and now a storm had taken him. The big man could not speak.

Leon looked up at the Christmas lights, whipped by the wind. John Blewett had helped him put them up the day before. Now his childhood friend was gone. That night haunted him for years.

The village was overwhelmed with grief. Slowly details emerged. The family members found out the men died trying to save two young girls and their mother.

The international press descended on Mousehole and the grieving families were hounded. A hell of a lot of money came in from fishermen in Scotland. People crammed notes into jam jars in shops, thinking about those families. £3 million. A lot of money for a poor community.

This fishing village had its heart torn out. Christmas should have been a magical time with everyone packed in the Ship but now there was just a sign in the pub window saying that Tom Bawcock's Eve, a traditional dinner, was cancelled. When John, Nigel and Leon were kids they entered the sculling race and competed like hell, all fired up. Then they organised Mousehole harbour sports for their own kids. They rigged up fifty-metre lanes for swimming races and had rowing-boat contests. The kids dived for plates, climbed the greasy pole and had pillow fights. Major Kelly from the Lobster Pot restaurant put up money for the prizes. After the sports they packed out the Ship, everyone taking the mickey out of each other.

All the pillars of the community had drowned. That was the end of the Mousehole Christmas lights. Only Leon was left.

Edwin Madron went up to the court inquiry in London with his father Jimmy. In the end Jimmy got so fed up that he got up and said: 'You're all talking a pile of bollocks!' Everybody was blaming each other. The locals blamed the coastguard for not calling the lifeboat earlier. *The Union Star*'s skipper was blamed for not accepting a tow before the weather became too bad. The coastguard blamed the tug. Jimmy Madron had already lost his brother in that accident off Plymouth when they were kids. His son Edwin was on the lifeboat crew too. He would have been on it that night, but he had been in Ireland. When he came back he said to the old boy Cyril Torrie, who lost his son: 'I don't want to bring it up, but what was it like that night?'

'You couldn't see a fucking hand in front of your face, because it was blowing so hard, and the sea being whipped up and pushing that in your face, your eyes, we nearly went over the cliffs.'

So all these people saying they saw lights steaming around. Fucking liars. Or that they were on the cliff watching the lifeboat. Lies! It was a hundred-mile-an-hour wind you couldn't look in front of your face.

Every anniversary inquiries are fielded from newspapers; the money shot in these touchy-feely days is a shot of the families crying. They have refused to take part. The first commemoration was on the twenty-fifth anniversary in 2006. They had coffee in the chapel. Many of them spoke about it to each other then for the first time. The way of fishing communities is not to grieve, but to see it as one of the blows that falls. That's the culture.

Sea stories speak to us; when the boat was lost, everyone lost someone. There have been five attempts to make a film. Sculptures were made. A folk song was written by singer Seth Lakeman, calling them 'heroes'. Some people felt uncomfortable with the word 'heroes' and pointed out that the Lizard lifeboat was heroic too. Part of their rail was missing; the boat's hull was wrecked underneath when they got back; the bilge keel was missing and the bilge was filling up. The boat was breaking up. How they survived that night, no one knows. Martin remembers listening to them talk about it in Cadgwith Cove Inn and seeing how their nerves were shredded. The next day there was still a hell of a sea running, but the bay was full of fishing boats, looking for wreckage. They couldn't bear to stay at home.

Edwin Madron, the last sea dog, became the harbour master at Mousehole. He oversaw the activities of the ten fishing boats there. He remembered the days of Mousehole as a thriving fishing port with 200 pilchard drifters, mackerel drifters, ring-netters and longliners. Forty-six years a fisherman, he carried on on his own and could be seen hauling his heavy gear around Newlyn quay. Nigel Brockman's son Neil became the coxswain of the Penlee lifeboat. Neil's own son joined the crew later. Jimmy's other sons still fish out of Newlyn. Baden is in Newlyn, married to Stacey's ex-wife. Shaun is still a fisherman. Stephen's widow stayed in Mousehole. So did Nigel's – at the end of the village. There a hard kernel of them remain, and see out the winter storms.

John Blewett's widow eventually met a farmer she liked from upcountry and asked Leon what he thought. Leon said settling down with him would be the best thing she could do because all the bums in society were pursuing her for her money. John Blewett left a 12-year-old son behind, Andrew. His dad had been second row, captain of the rugby team. Andrew, also known as 'Blewey', was a good rugby player too, toured the Welsh clubs and pubs from Cardiff to Llanelli.

A few years after the 1981 disaster, the Big Bang in the City happened and Mousehole flooded with yuppies. A London estate agent came down and went through twelve houses in a couple of years. He would buy a house, gut it, do it up, put his son in it, then another one for his daughter and another for himself.

Blewey didn't like emmets at first, but he was a gifted carpenter and he was in demand to do up all Mousehole's pilchard cellars and net lofts. The Old Pilchard Press in Mousehole is rented out now for £1,250 a week. He did up the quayside house of a London-based CEO, and found he was a beauty – he was down-to-earth, happy to stand and have a pint and a chat with Blewey. Blewey is always down the Ship Inn in Mousehole, where harried mums in Boden Breton tops ask his daughter to babysit. Blewey has the most piercing sea-blue, quartz eyes; lodged above the blackest beard, those eyes look right through you. Blewey stands outside to smoke and watches his kids jump off the harbour wall in their wetsuits. He loves waking up in the morning and looking out.

'If there's something better, let's go and do that,' he says.

Only in winter does Mousehole belong to the old locals again. Leon walks out and he can't see a single light on around the harbour. All he can hear is the noise of the waves battering the south quay, the wall of granite boulders protecting them from the fury of the ocean.

4

Martin borrowed £200 off Nick Howell to buy the Madrons' net. It took three men to put it in the trailer, which gave him pause for thought: if a net this size was full of fish the weight of the catch would be too much for his small boat, and pull it down. So he spent weeks making the net smaller, taking a panel off and sewing it back together. It was a hundred fathoms long and ten deep; twice as deep as a church tower, half the size of a football pitch. Still a huge net, but it wouldn't – quite – sink his boat.

His first large catch was half a mile off the Penzance swimming pool. He spotted pilchards this time, not mullet: 200 quid's worth, then more off Falmouth the same day. He did a little dance to see the threshing mass of silver gleaming in his net. A shoal rushing on the surface. These were the moments Martin loved. It was heartening for Sally too, to see Martin come home happy with bigger catches. More money was coming in. He would go out at 5 or 6 p.m. in the early evening and be the only one in Mount's Bay.

Ever since they'd first met, Nick had been talking to Martin about his dream: reviving the pilchard fishery. Newlyn was built on pilchards; they shipped them to Italy. As long as there were Roman Catholic countries that ate fish on Fridays, the blood of pilchards would roll down the streets of Newlyn. The silvery fish had once been the backbone of the Cornish economy. The shoals were still out there in the sea but no one had brought them in since the 1950s. This was simply a cultural shift: everyone thought of them as cheap student food in horrid peel-back tins of tomato gloop, best left on the bottom shelf of Spar. By 1995 Cornish landings were a mere seven tons a year. Nick knew stocks were a thousand times that. And there was no quota on pilchards. They were oily fish; good for the heart. The challenge ate away at him.

The first threat was the bloody new EU regs. They said you can't use wooden presses and barrels anymore. You had to use stainless steel. Fuck that, Nick thought. He got round it by turning the Old Pilchard Works into a working museum. It told the story of pilchard fishing in Cornwall. His team salt-cured the silvery, oily fish in wooden barrels. Prince Charles came and opened it. No one wanted grilled pilchards. But sales of grilled sardines were taking off in Cornwall. People told Nick they tasted like the sardines they'd barbecued on holiday in France or Spain. It gave Nick an idea. In 1997, when Marks & Spencer asked him to supply some French sardines, he gave them Cornish pilchards instead. The pilchard is, after all, an adult sardine. The buyer loved the taste and asked for more. If Nick could rebrand the fish as Cornish sardines, then the pilchard fishery might revive.

Martin was excited about Nick's plan. Pilchards weren't found in Devon or Wales – only Cornwall. Martin loved going after pilchards. Large shoals had in the past congregated off Wolf Rock, but regular catches had always been a problem. Martin tried everything, from using lights under the boat to drift nets.

Martin had found his calling – pilchards were traditionally caught in ring nets – and his market. He landed his catches at Nick's pilchard works. By 1998 the price for pilchards had risen tenfold. Everything suddenly made sense. Within a few years Cornish landings of pilchards had soared from seven to 200 tons. Martin asked the bank manageress for a loan to buy a new boat, a thirty-footer called the *Penrose*. He told her he could catch five tons, and sell one ton to five different buyers, as a way of reassuring her he had some marketing contacts. She agreed. He would carry the catch on deck rather than down below. He would prefer a pukka fibreglass boat with a fish hold to store fish down below, but the *Penrose* was all he could afford. He'd put a hauler on a high metal bracket to bring the net in.

In the Newlyn pubs they joked that Nutty, from across the bay in Cadgwith, was trying to show them how to catch their own fish. But behind the banter everyone was intrigued to see if Martin would succeed or fail. Everyone was watching him.

The winter of 1998 was savage. There were fierce gales. Sometimes Martin couldn't get out. Sometimes, when he could, he couldn't find the pilchards. Empty nets meant that his crew didn't get paid, and ultimately that he wouldn't be able to keep them on. So he moved out of Newlyn to look for shoals in Falmouth over the winter. Then he looked in Mevagissey but he still couldn't find them. Summer was difficult too: he tried crabbing again, but his crab pots were towed away by scallopers who dredged up the sandy seabed where he'd laid them.

Martin didn't have much money left after paying off the bank, insurance, diesel, food. The *Penrose* didn't have much going for her either: no sonar. Without sonar, the ocean seemed endless. Going out to find shoals like that was just rolling dice.

Sonar was expensive, a one-grand piece of kit. But he took the plunge. Now Martin could see the swivels on the screen where the sonar had detected individual fish forty-five fathoms down. He loved the eerie echo sound it made, like in a submarine war movie.

When he came back to Newlyn in autumn 1999 he found much larger shoals of sardines than before. One was enormous, eleven fathoms deep. He closed the net in its midst so it was bursting with fish. Then the whole shoal dived down to escape, pulling the net and head-rope down so deep the boat began to lean over. What the hell was he going to do? The catch was worth thousands. The tension suddenly gave. The force of the fish split the net from top to bottom. He was encircling too many fish at a time. He stayed home and carefully stitched up the net.

A few weeks later, just before Christmas, nets mended, Martin waited on the quay. It was already dark. The wind was due to pick up at midnight. Martin's crewman had not shown up and he was two hours behind schedule. A couple of local lads, Carl and Patch, offered to help out.

In high spirits they steamed out three miles south-east of Mousehole. He felt proud, showing these young men how his sonar worked, guiding them through his ring-netting technique. They found a shoal of sardines, shot the net, then he steered the boat slowly around the shoal, in a full circle, to pick up the net end. Once the shoal was surrounded, they swiftly heaved in the rope threaded along the sea bottom, trapping the fish. Patch and Carl were fit lads who played rugby for the local Pirates and only took an hour to haul the catch on board, scooping them up using a small brail net. They took on as many as the boat could carry, then let the rest go. A powerful torch and the deck lights played over hundreds of silver fish, fighting to get out. They watched them and caught their breath.

Martin whooped out loud at such a large catch – 500 stone, worth about a grand. He felt like dancing. The *Penrose* had no hold, so the pilchards were left sloshing around on deck, a seething mass of oily silver.

There was a lot of tide that night. The three-feet waves in Mount's Bay were twice that size off the Lizard. The wind groaned. Dark waves reared up and buried the bow. Spray lashed the wheelhouse. Martin kept the bilge pump going. He started to worry: even big boats flounder on the Lizard when the waves swamp you. They can hurl you against the hauler, or the derrick can list. A few years back two Cadgwith fishermen set sail from Helford and disappeared. Never found the bodies. The village grieved for weeks.

In bad conditions, seawater drains overboard through a boat's scuppers. Because he had to keep his catch on deck, Martin had put mesh over the *Penrose*'s scuppers to stop the fish slipping through. The pilchards were thumping round in boxes, jumping about all over the place, sloshing back and forth, as the spray came over. The boat felt top heavy and unstable. The radio wasn't working, so Martin felt around for his mobile phone. He called Nick to say they were on their way.

'We have a good catch on board but the wind is starting to freshen. We're three mile off Mousehole.'

Nick knew Martin well enough to detect a quaver in his voice. Martin didn't deliver information without a joke; this time he had blurted out his position. Nick phoned Falmouth coastguard to put them on standby. The coastguard called Martin's mobile. Martin thanked them.

'We are three people on board, Carl, Patch and me.'

He gave them the course and position, told them to call back in half an hour.

Ten minutes later the *Penrose* was in serious trouble. Water kept coming over the bow. He put the deck lights on and saw hundreds of fish rolling round the aft deck, in five inches of water, and it wasn't clearing. The fish had choked the scuppers. She was rolling. He realised the catch was too big and the boat wasn't seaworthy enough to carry it.

He did the maths. There was a ton of water on deck with 500 stone of fish. The net weighed another ton, the winch half a ton and the powerblock was 300 pounds up in the air. Too heavy.

'We're in the shit. I think we're going to sink,' he told the guys. 'I don't know what to do here. The waves are too big to open her up.'

'Don't worry about it,' said Patch. 'Open her up.'

He should have done it, should have opened the engine up flat out, shut the door, got the bow up and hoped that water didn't come over. But he eased her down.

The others put life jackets on and went out of the wheelhouse. The boxes slid over to port. Carl jumped into the sea. Patch threw the life raft over the side, yanked the painter so it popped out of its container and inflated with a bang. Then he jumped overboard too.

Martin was left alone in the wheelhouse. The sea level seemed to go down. He could feel his Barbour, mobile and sonar were soaked through; the boat was going over slowly onto her side. He couldn't tell if it would keep going down. He knocked the engine out of gear, climbed out of the wheelhouse and waded towards the stern. The mast sliced down, and with it the wires and stays that kept it up. He ducked underneath so his life jacket didn't snag. Then he clambered to the stern to jump, but found the net's floats were up. To clear the net he had to hurl himself far out into the sea. The shock of freezing water hit him. It was a bitter, icy December, just before midnight. It was very dark, cloudy; miles away in the distance car headlights glinted, on their way from Mousehole to Penzance.

The *Penrose* lay on her side, engine still running. Ford, 120 horsepower.

As he swam away, he felt a rope tug him back. It was caught round his foot, and as he turned in the water it skidded up his leg and wound round his groin. He yelled out to the others. He couldn't get it off. Fuck. He heard the sound of the exhaust, running chom chom chom chom. Screams cut the air above his head – he glimpsed gulls wheeling in the dark. The deck lights were still on underwater. It was an eerie light. Sachets of strong fluorescent dye had spilled into the sea round the wheel-house. The water was full of weird colours, lights and noise, like a rave.

There was too much going on. Boxes were floating all over the place. Soon the sea would flood the wheelhouse and she'd be gone.

Finally he got a grip of the rope, worked his fingers under the coil in the water, tore the rope off his leg and swam to the life raft, hyperventilating, trying to breathe. They pulled him in.

'Clap your hands together,' Patch said.

He must be panicking.

'Exhale. Back in again.'

The deck lights played over Patch's and Carl's shocked faces. The life raft drifted with the wind until they couldn't see the *Penrose* anymore. No one looked back for it either. They were wet through, in a biting wind, as six-feet waves hounded them. They let off a flare, watched it fizzle and die.

There was a little bit of food aboard, a knife, some more flares, water. They put a drogue out, a sea anchor, which swung the raft round away from the wind. They waited for forty minutes. Nothing.

'That was a bit of a lash-up.'

'What happened there?'

But no one felt like talking. They could see St Michael's Mount silhouetted in the distance.

Martin's mobile was lying on the seabed when the coastguard called it. That was not a good sign so they launched. They held for the coordinates Martin last gave, looked at the charts. The Culdrose search-and-rescue helicopter went up and raked the sea. In the darkness the Penlee lifeboat made out the flickering light of a hand-held flare.

The huddled crew in the life raft heard the hum of the lifeboat. It sounded like it was not far off in the dark so they let off their last flare. Minutes later they saw the wheelhouse lights of the *Penrose* sink down below, impossibly far. Then with a roar the lifeboat drew alongside; strong arms heaved them in, slippery and wet, like newborn calves.

'Blow, Patch, I wondered where you was,' they joked. Patch was a lifeboat volunteer that night. 'We'd been on standby for half an hour and you were out here already.'

Rotor blades whomped above, as the Culdrose helicopter banked and flew back to shore. They steamed into the fishermen's mission in Newlyn for dry clothes and a hot cup of tea. Nothing tasted as good as those custard creams. Martin was given a lift home. Sally told a local paper that having him home alive was the best Christmas present ever. Toots was in the bath when she heard. Whenever there was news about her dad she seemed to be in the bath. He's sunk his boat, her mum said.

Everyone at school asked because it had been on the news. Local divers filmed the wreck – it was creepy for Martin to see fish swimming round the boat, rather than stowed in it.

He didn't fish for six months.

Billy Stevenson saw Nutty Noah sink that night. He watched the lights and flares from his hilltop window, overlooking Mount's Bay. For decades Billy ruled Newlyn. Now he stares out from his great living-room window with powerful binoculars, watchful of all the boats in the quay, who's in, who's at sea. He can identify the fishermen by the way they walk. He phones the boats up and asks them what they've caught.

Nutty Noah was the first one to go catching pilchards with a ring net. Billy said to him one day down the quay: 'You wait, you'll have nothing with that. The Madron family tried it eight years ago and didn't catch a fish.' That afternoon he walked past the ministry and they told him Noah had caught 400 stone. He grunted but said nothing. He pushed his glasses up his nose and walked on with his hands behind his back, deep in thought. Pilchards were always there, but no one could catch them. Stevenson boats came in at night and said they'd been steaming through pilchards for an hour, on top of the water.

Billy likes Nutty Noah, but not Nick Howell. His story is that Nick got a grant for a pilchards' museum but then sold it off as flats and retired early from Newlyn. Nick says he only got a grant for £26,000 from the EU to change the processing museum in 1992. Years later in 2007 Nick invested huge amounts of his own money converting the flats. He sold 4 but still owns 6 of them. Billy refuses to call pilchards 'Cornish sardines'. None of the old recipes in his wife's book calls them Cornish sardines. They are pilchards.

The *Penrose* was sunk in ten fathoms of water. Billy offered to help lift the boat from the seabed. He sent out his biggest beam trawler, the *Cornishman*. They tried several times to haul it up with a big crane, but watched as it came up stern first then broke in half and slipped back and disappeared below. The rest of it was just hanging there. The great Madron net lay in tatters on the seabed.

Billy took the plate from the *Penrose* and added it to his dusty, private museum that no one ever sees: room upon room of an old primary school, filled with old engines, anchors and propellers brought up from the sea. Like Miss Havisham, Billy surrounds himself with decaying relics of the past.

Billy knows the Stevensons have never had a good name in Newlyn. They said he was Victorian in the way he treated people. If you get on in Cornwall they would rather a stranger took your place. It's definitely a Newlyn thing. So he never sets foot in the Fishermen's Mission, but visits David Baron's paper shop instead and asks: 'Who's dead today, David?'

In his living room too he hoards artefacts: boats hammered from Newlyn copper, gourds trawled up from the ocean floor, port and starboard lights, a toy wheelhouse his six grandsons played in, and a vintage brass ship's telegraph that used to ding out commands from the pilot's wheelhouse to the boat's engine below: full, half, slow, stop, finish with engines.

'The trouble with you, Dad, is you've lived too long,' his daughter tells him. He's 86 and has seen the fishing ports of Lowestoft, Grimsby and Hull go down, but Newlyn survive.

Billy's grandfather was a pilchard fisherman. Billy's father built the Stevenson trawler fleet up after the Second World War. Billy went to sea after he left school at 15, in 1943. His father wouldn't let him go out until after six o'clock, so for years he'd take his boat, the *Girl Sybil*, out on a black night, with just a compass and local knowledge, using the natural light of the moon. He got on well with the fishermen, because he knew what he was talking about.

If they said it was a hard living, he asked: 'Why don't the crew ever lose any weight?' If they were superstitious about eating a pasty at sea, he said: 'Eat the buggers on the pier.' One man didn't want to sail on Fridays because a Brixham boat was lost on a Friday. 'Go at five past midnight,' Billy told him. He told the fishermen he kept a bit of wood in his pocket to touch for luck but they thought he was taking the piss, said he was a fucking liar. 'I'm going out today, we'll have a nice bit of fish,' one skipper said, as he was about to set sail. But Billy would never tempt fate. 'Man proposes, God disposes,' Billy replied.

The Stevensons only ever lost one boat, the *Boy David* off Tater Du, but the crew were saved. Twenty years ago they would have gone out in the bad weather, because their firm was built up on a lot of good men. It's different now; the crew rely on instruments, not skill. One skipper rang him up complaining, 'My Decca's packed up. Can I have a tow in?'

'You know where you are,' Billy said. 'You can see the lighthouse on Wolf Rock. Just steer in.'

The skipper didn't understand. Modern technology ruined fishing. When echo-sounders came in, they fished the dogfish out the water in two years.

Stanhope Forbes, the artist, complained when the boats became motorised, because the pilchard drifters with sails were prettier to paint. Billy thought Forbes was a nuisance. Once as a boy Billy snuck up behind him to peek at his canvas. Forbes turned and daubed him in the face with his brush.

At night Billy stands in the dark and can see from the lights on the quay how many men are living on the boats. Men who've been in the pub or men from the Lizard half the time.

Billy remembers when the Penlee lifeboat, *Solomon Browne*, was lost. They rang him because his wife was related to one of the crew, and the lifeboat coxswain, Trevelyan, was a skipper on one of Billy's boats. It should never have happened. The American in the helicopter that night didn't know the shore.

The lifeboat they've got now is a beauty. Billy's been aboard her once. Patch is a good lifeboat skipper.

Billy and Tony, his younger brother, were different. Tony was favoured by their father. Billy was the more formidable one, always working away at the boats. Tony was the showman and athlete; he played rugby, he swam for the county. He focused on the fish auction, the property, all the stuff onshore. Billy and Tony had rows.

Billy never went to meetings; he was busy down the quay. Tony was fed up with going to them, talking about methods of fishing with a group of fishermen: what did that ever achieve? So Tony looked round the office and saw Elizabeth, Billy's eldest daughter, in her 20s. She started coming in the office at 17, in her school holidays, after she had finished her O levels. She did a bit of bookkeeping. Tony walked over to where she sat next to her grandfather on the main desk.

'Go to this meeting in Plymouth for me,' Tony said.

She'd never been to a meeting before. She took her place round a mahogany table of white-haired men. She started to explain why Tony wasn't there.

'Tony's resigned,' one replied, gruffly. 'You're on the committee now.'

That was that. She was part of the South-West Fish Producers' Association. Elizabeth was the eldest of the next generation; she had two sisters. In family businesses, you might not want to do a particular job, but you ended up doing it. Billy loved messing around with boats. You couldn't tear him away from the quay. Elizabeth lived fishing, being Billy's daughter.

When she started work they could go to sea with any boat, any size, any engine, catch any fish when they wanted and land it where they wanted. There was no paperwork. Every time a new regulation came in, she talked the skippers through it. The fishermen started to seek her out. They knew she was interested and they started coming to her with their other problems: health problems, money problems, wife problems, children problems. She'd find a solicitor for them. If they didn't have a doctor, she'd register them at a surgery. She used to do all sorts of things.

One day she bumped into someone in Penzance.

'Hello Elizabeth,' he said. She didn't have a clue who he was.

'You don't recognise me do you?'

'No, I'm sorry, I recognise the face but I can't put a name to it.'

'You helped me get my house. My mum came and saw you, I was having problems with my mortgage, declaring my income and things. You sorted it. And I shall always be so grateful.'

She used to go to the bank and come back with £50,000 in cash in a carrier bag. It was handed out in the office to the skippers, then they would pay the crew in the pub. Their money used to be put in little rubber bands, wrapped round a sheet of paper. Sometimes the wife turned up in the pub. Some men had already split the wad and stashed half in their back pocket. The wives found it. They had to run things. They looked after the children, the finances, took the kids to school, chose the school, did the after-school clubs and parent-teacher evenings. It was the same with Elizabeth. Her husband, Sam Lambourn, was hardly involved with the schooling of their three kids because he was at sea. Even when the kids were teenagers. She had the help of a nanny, but when Elizabeth walked in the door, the nanny left. She did the kids' breakfasts, took them to school, went to work and came home.

Nobody paid her to be on committees, but she got feedback on the gear, the quotas, the seabed, the hours they were working, any difficulties they had, trouble with the weather. She'd never had any problem talking to crews, skippers, or other owners down the quay. It was easier than talking to a gathering of mums. She had no problem shouting at one of them. They didn't have a problem talking to her. She was treated with respect, part of the group.

'What's it like being in a man's world?' people asked her.

'It's my world.'

She knew a lot about it and liked it; it tested her. It wasn't something she would have gone into if she had wanted to make shedloads of money and have an easy life.

She'd been on every committee in the world, ending up as secretary of the Cornish Fish Producers' Organisation and chair and president of the National Federation of Fishermen's Organisations. She represented the UK fishing industry in Brussels a couple of times at the fisheries councils.

Fishing is a hard life. Her husband worked hard in Mount's Bay with his purpose-built catamaran, *Lyonesse*. In her holidays their son took his boat out and came back aching. They did it because it was a challenge. When the boat was laid up they were dying to get out there again. It was the freedom and space. They didn't want to be in an office environment. It was the camaraderie. Some crews seemed more worried about each other than their family at home.

Billy'd been a big influence in her life; she was a chip off the old block. 'What would he do?' she often thought when problems came up. But she didn't always do it. She read up on a lot of books and articles on family businesses and found that not many lasted as long as the Stevensons' had. It was much easier to fall out with your sister, father, aunt or uncle if they were your business partner too. There were eight family members trying to agree. She lived near her sisters, worked with them, had children at the same school and went to lots of things together.

Elizabeth had this reputation of being like her dad: hard, harsh, and a bit Victorian. She had very fond memories of going abroad with Billy and buying the Dutch beamers.

'Dad, could we have one like that one day?' she said, pointing at one with an old-fashioned slanting wheelhouse and galley windows. They bought her the next year. When she watched a couple of their boats being broken up at the end of their sea lives, she cried. She cried when her daughter went away to university, even though it was her third year there. When she saw her children perform in the school play, when her daughter did a couple of solos, she cried.

Everything she had, she'd got from the Stevenson family.

Looking down through his binoculars at the fishing fleet below, Billy's mind is pretty switched on. Above him, higher up on Chywoone Hill, Elizabeth's own glass-walled living room looks over his house at the quay. She too can see what the men are doing. She also knows all the boats. She can recognise quite a few of them by their lights in the dark. At teatime, she will think why hasn't so-and-so sailed? Do you know so-and-so is coming in? There are people living on boats, but not Stevenson boats.

She still gets all choked up when she hears sunburnt men sing in harmony, in pubs from St Just to Cadgwith:

I've stood on Cape Cornwall in the sun's evening glow,
On Chywoone Hill at Newlyn to watch the fishing fleets go.

She liked to go down the quay when there was a huge amount of activity: lights blazing down on red boxes of shimmering silver fish, being packed in slush ice. Cars and fish lorries with engines steaming. It meant that money was changing hands. People from Grimsby, Padstow, St Ives, Falmouth. When she looked on the automatic system, AIS, that tracked all the boats, she saw there were loads of French boats working. Stevo beamers went a long way off: they were not allowed to fish in Mount's Bay, inside the twelve-mile limit. Sometimes a Newlyn hothead met a French hothead and gear was cut away.

She didn't know how she did it. She'd always worked hard. It wasn't given to her. Somehow she kept her head above water, kept on going.

Then in 2002 the shit hit the fan.

Everyone had been grumbling about the quotas. High-value fish like cod, hake and anglerfish were heavily restricted by EU quotas, aimed at preserving stocks. But more hake than they were allowed came up in the nets. What were they going to do, throw them back in? The fish were already dead. Some skippers and owners labelled them as 'turbot' or a different, cheaper species on the landing documents, then sold them at auction at their real value anyway. Over a while it became a conspiracy, with 'black fish' coming in under falsified documents. Elizabeth always knew they were doing wrong. But it was going on across the whole country, so she reckoned everyone knew about it – from the minister at DEFRA, to everybody else.

It wasn't money laundering. It wasn't really false accounting, except it might be *called* false accounting. Everything was paid and done on the paperwork. It was all going reasonably well.

Then the Marine Fish Authority launched an investigation, using a lot of sample boats. Some were Stevenson boats, some were owned by other skippers. They uncovered the scam. It became a notorious scandal. The prosecutor said the 'black fish' conspiracy had been going on for years. Six Newlyn owners and skippers had fiddled their books.

The court case began in 2002. Elizabeth sat at the back with her legal team. Truro court flooded with natural light. She could hear seagulls circling above. She was to lead the case through the courts. It went on for six and a half years. It was terrible.

The Stevensons did not own most of the boats; they were agents for them. They decided to plead guilty. It was a test case, designed to send shockwaves out to all the ports and fish auctions in the country. Right up to Scotland. Two days after they put in their plea, Elizabeth looked out the back office window. She saw a police car come down the quay. It had a black grill over the front.

'Somebody's in trouble!' she said.

She turned around a few minutes later and they all started coming in the door. Flashing warrants and Christ knows what. Everyone had to stop working.

Open doors. Open safes and cupboards. Then there was a phone call.

'Elizabeth, it's your husband.'

'Tell him I'm busy.'

'He says you have to come to the phone.'

She came to the phone.

'You've got to come home now.'

'I can't Sam. I just can't.'

'*I want you home here now.*'

'I can't. Please.'

'Bloody come home now. I've got police swarming all over the house.'

They took their computers. They went through their drawers. They went through the whole house. They filmed everything. These cops were from the asset recovery agency, who usually seized ill-gotten gains from terrorists and international drug barons. They were going for £4.5 million. It was a hell of a lot. When the Stevensons had pleaded guilty to that kind of offence, they were then deemed to be living off the proceeds of crime. Normally the prosecution had to prove they were guilty. Now her family had to prove they were innocent. In the very early stages the barrister asked her: 'If you could settle out of court, what would you write the cheque out for?'

'£600,000,' she said.

She thought that was fair. It went on for three more years with the asset recovery agency.

Billy shrugged off the court case. They were caught out a couple of times with other people's fish, from seven Newlyn boats who used their stores. 'The Marine Fish Authority or whatever they call themselves are trying to ruin Newlyn!' he said. He was still annoyed years later. It took its toll on him.

He had a coronary bypass. It was on the cards that he might die. So he retired. Tony took over. Billy would watch the fleet come in at night. *Why wasn't Tony down the quay? He'll let the business fall to pieces.* Billy's father never let Billy leave the quay, if boats were coming in. Billy once wanted to take his future wife Enid to the pictures to see *Sands of Iwo Jima* with John Wayne, but his father kept him in the harbour so late that the film had long finished by the time he ran to the cinema to meet her. Eventually, one night Billy went outside, to the house next door where Tony lived. Billy banged on the door, hammered on the window.

'Go down the pier!' he shouted in the darkness. They fell out, didn't speak for years. Tony fell ill and died in 2010.

On the last day in court, January 2009, Elizabeth made the trek to Exeter for the verdicts and sentencing. The fishermen had not defended themselves. Some of the defendants were elderly part-owners of family boats. One old lady was excused because she knew nothing about the boat. The judge put a ban on reporting. The prosecuting silk stood up: 'It is clear from the scale of the offending that this was a system operated at this auction house and can only have occurred with the active conniv-ance of the auctioneer and other staff acting for W. Stevenson & Sons. There is a lot of money to be made from selling black fish . . . The Stevensons were at the heart of the case. These were deliberate and well-organised deceptions.'

Doreen Hicks, 82, had taken tranquillisers to help her through the trial. The judge sentenced her and the other thirteen defend-ants to pay £200,000 or face jail. Doreen sobbed openly in court. The Stevensons had to pay far more. Elizabeth wrote out a cheque for a million quid after costs were added. It was a hell of a lot of money.

Elizabeth walked outside the crown court, into the fresh air. She pulled a thin cardie over her floral dress; in matching pearl necklace and earrings she faced the press. The reporting ban had been lifted. She felt her hands shaking later as she read the *Daily Telegraph* headline which called the Stevensons 'fishing pirates of Newlyn caught in law's net'. It described the defendants as 'the self-serving fiddlers of Newlyn who conspired to wipe out our marine resources for private gain'. She paced the room, clenched her jaw and flung the paper in the bin.

It was a terrible time; a real low. But Elizabeth fought back. Her nickname wasn't the 'Queen of Fish' for nothing. The Stevenson family had been in business since the nineteenth century. In Newlyn the large fleet of beam trawlers they'd built up dominated the port. She wasn't going to let this episode blot what she'd achieved. After all, she'd met the royal family at Holyrood Palace. Her husband had been invited to St James's Palace. She'd gone to 10 Downing Street to meet Tony Blair; she was pregnant at the same time as Cherie. In 2000 she met Margaret Thatcher. The highlight was during her tenure as president of the National Federation of Fishermen's Organisations when she'd laid the wreath at the Cenotaph as part of the Remembrance Day ceremony with the Queen and chiefs of staff.

The many fishermen in the pubs on the quay who were on the Stevenson payroll closed ranks behind her. Her legal team had argued fiercely that a huge additional fine, on top of the £1 million, would hit Newlyn's community too hard. So the judge only charged them an additional £25. In the Star and the Swordfish the locals were divided. Some thought the million pound fine served them right. Others thought Christ, there but for the grace of God go I: the Stevensons took the rap while hundreds got away with it. DEFRA was trying to crush Newlyn, the final Cornish outpost, the last stinky, oily fishing village, before the second-homers moved in. As someone said: if too much hake and cod came up dead in the nets, what were fishermen supposed to do – give them the kiss of life?

It put the Stevenson business back. They couldn't invest in anything for six years. The family accepted £920,000 to decommission two beam trawlers, including the 283-ton *Daisy Christiane*; a measure to protect sole, which was at risk in the Western channel. Elizabeth cried when she watched the trawlers being broken up. The hundred-year-old Newlyn fleet went down from twenty-eight boats to thirteen beam trawlers.

'The boats are rusting up,' people complained down the pub. 'If that was an airline would you fly on it?'

Elizabeth has heard the complaints but she also knows they cost an awful lot of money to replace. They are seaworthy. They are safety compliant. She couldn't live with herself if they weren't. They have a refit programme.

'Newlyn will go down!' Billy would shout bitterly across the room, shaking his head, clenching his fist. 'Newlyn will be ruined anyway in a few years' time.'

They postponed plans for buildings that the community wanted, blaming crippling fuel costs, the unworkable quotas. Billy railed against the crabbers who fished with no restrictions. His beamers needed licences and four nimble hands to shoot the gear, but crabbers used cheap labour from Eastern Europe, who live on the boat. Anyone can crew on a bloody crabber!

Elizabeth listened to Billy fume about Newlyn, but she was not sure he believed it. In 2011 three of their beamers landed record catches of over £51,000. She thinks Newlyn will be there long after she's gone and thinks of her son, who's at Cardiff University, going out in his punt mackereling, squidding. He likes fishing. In 2015 she is building her own fleet. At the centre of the black-fish scandal was hake. Elizabeth had a company with its own hake quota, so she and Sam decided to catch it themselves. They bought a hake netter. She has been looking for another one. She wasn't a Stevensons' partner anymore, just a director. Family still holds the shares.

Elizabeth often gives talks to the Women's Institute. She is unrepentant. 'Imagine if a photographer took thirty brilliant shots and a law tells him he can only sell ten,' she says. 'If he calls them posters instead of photos on his invoice he can sell twenty. Wouldn't you be tempted to sell them as something else if you could get away with it? Rather than throw them back or burn them?'

Newlyn is a law unto itself.

'You never seen old bill down Newlyn,' fishermen say as they openly puff on spliffs on the sunlit benches outside the Swordfish, where that policeman was once jettisoned through the window. The Stevensons' disregard of the quotas echoes an age-old Cornish resentment of laws imposed from far away and their rebellion against Westminster's taxes. The far west of Cornwall has long been the kernel of resistance, where people are the most defiant and lawless. The history of Newlyn and Penzance is defined by one word: cussedness. They stubbornly did things their own way. At the beginning of the Middle Ages, London's power waned and Cornwall was left to its own devices, living in feudal anarchy.

For hundreds of years the English saw the Cornish as rebellious. They didn't answer to Parliament. They fought against Parliament in the English Civil War. The king granted them independence, as the Duchy of Cornwall, so they backed the Crown. It became a Royalist stronghold, like Wales. They bitterly resented the English. A Cornish army swept over the Tamar and stormed Bristol. They took Lyme Regis on the coast. The Roundheads were unnerved. They had to crush the Cornish. So the Earl of Essex invaded Cornwall with 7,000 men.

The Cornish country people were 'more bloody than the enemy'. They picked off any stragglers and slit their throats. They hid all food and provisions from Essex's men, who had to march thirty miles on a piece of bread. When Cornish forces hit back, Essex fled on horseback for the coast. His infantry, bogged down in mud and rain, surrendered. In Lostwithiel the soldiers were mobbed. Violent local women stripped them, tore their boots off and left them naked under hedges. The terrified men fled for the Tamar, only feeling safe when they had escaped from Cornwall. Of the original 7,000, only 1,000 made it back. Panic spread amongst Cromwell's Parliamentary Roundheads that 'Hellish Cornwall' was full of pagans and heathens.

It was a towering Royalist victory.

They were led by a classic Cornish rogue. Sir Richard Grenville was a penniless, black-eyed, war-hardened soldier and ex-convict who deserted Cromwell's ranks to further his own ends. He rallied ordinary Cornish locals, including soldiers who were women dressed as men. He understood they harboured a dream of independence. He extorted cash from the Anglicised gentry and refused the king's orders to push east into England. Instead he ordered his men to garrison the bridges across the Tamar and keep 'all foreign troops out of Cornwall'. He wanted to establish a semi-independent Cornish state, led by the Prince of Wales, who was also the Duke of Cornwall. His troops even blocked other Royalists from retreating back into Cornwall. The dream of Cornish autonomy was near.

But Grenville's enemies closed in. He'd starved chained Parliamentary prisoners to death. As a deserter there was a bounty on his head. He was imprisoned in St Michael's Mount. Then fled abroad with the Prince of Wales.

Many Cornish still harbour that dream of independence. They don't follow English rules.

Martin didn't know what to do. Should he keep experimenting with ring nets or give up? The *Penrose* had broken into pieces; the monster Madron net lay in tatters on the seabed. He asked Sally to remortgage the house and borrow eighty grand to buy a new boat. She wasn't keen. Nothing seemed to be going right for him.

'I need a bigger boat for safety,' Martin insisted.

That made her think.

With the mortgage money and £26,000 from the insurance, he went to Ireland and bought a forty-feet red steel boat, the *Prevail*.

His daughter, Toots, came with him to Ireland. Four days off school is brilliant when you're 13. When they got there they found the prop shaft was broken. The new one took three weeks to arrive. They were stuck without a car. Toots was just about into drinking beer. She didn't have any friends there so kicked around the harbour with her dad. She helped sand down and varnish all the wood on the inside of the wheelhouse. She cleaned too, because it was in a right state down in the galley and down below. She didn't want to live in a pit.

The return trip to Newlyn took twenty-seven hours. Toots felt seasick. Martin used a bit of fishing net and rigged it up for her across the front of the boat to make a hammock. She climbed in, all cuddled up with her knees up, arms clamped to her ribs. In there with her Walkman on full, the tunes went boom boom through the sea for hours. Through one eye she saw dolphins breaking the surface at the front of the boat. She counted nine at one point.

They decked the *Prevail* out with flags as they drew close to Newlyn harbour. Through the gaps in the harbour wall they saw a couple of people on the main quay wave at them. Martin steered past them to the smaller, new quay and made out a distant cluster of twenty or so of their family and friends all waving madly. Toots and Martin stood in the bow, arms round each other. They could hear Sally and everyone cheering and whooping. It was six weeks since Sally had seen them. Now at last, here they were, back safe and sound, in a forty-feet, gleaming new boat. What a welcome. That was the first time Toots cried from happiness. It was amazing. They docked and everybody clambered aboard.

Martin kept fighting. He scoured the sea first for whelks, then porbeagle sharks. Porbeagles are playful; they roll and wrap themselves in long kelp fronds near the surface; they prod driftwood and floats. A canny fisherman can find fish without an echo-sounder, by knowing about plankton, predators, tides and the way the wind scatters over the sea surface. Martin knew that mackerel and pilchards fed on plankton and that the sharks followed the mackerel. He baited up 300 hooks and let them out on a six-mile line. He waited. He landed a good catch of porbeagle sharks. Using a hooked gaff pole, he heaved them on deck. He ran his hand over their gleaming mottled grey bodies. Large black eyes looked at him from snouts that tapered to a point. Their skin was soft; their bodies were huge and spindle-shaped. God's creatures, he thought, stroking their backs, as their tails thumped. Blood swilled over the deck. He landed seventy to Newlyn fish market, then another fifty the week later. It made him seven grand. That night, back in Cadgwith, pints flowed in the Cove Inn. His catch was written up in the local paper. An environmental group, the Shark Trust, called it disastrous, claimed that he was threatening the species and wrecking the ecosystem. He even received a death threat.

So he went back to pilchards.

As time passed, it became more difficult for him to find the crew. There weren't so many young men going into fishing anymore. The sons of Newlyn skippers were at college, or upcountry. The experienced crew were trying to get aboard the bigger trawlers where the money was.

He'd liked having Toots around. He never saw her now; she was always upcountry in Truro, finishing her first year of an art and design course. The other students were from Falmouth, Truro and all over. One was from the Scillies. None were from the Lizard. Learning about computers was so important for your career, they said. They were going to use the course to go to universities upcountry, or go to Falmouth. It was a real challenge making it in for the 9 a.m. starts.

'Oh my God, you didn't?' her mates down Cadgwith shrieked, when she told them she'd had to draw a wrinkly old man in life-drawing class. Stuck in a classroom, listening to the boring voice of her chubby, bearded computing tutor with glasses, her eyelids felt heavy. She missed the sea. She thought about her dad out there, with different crew members all the time. Then it was graphics class. That was OK. Textiles, painting, pottery were alright. Another tutor, Aaron, was from Penzance. All he wanted to hear about was the Swordfish, about fishing with her dad. He asked her what she caught. Her best trip she caught 200 mackerel, put them in fish boxes and sold them to people at school. She was 8. Her early memories were on the beach, swimming or going out in the punt. They went out off the cove, towards Kennack or the lifeboat station. Sometimes Toots and her sister tied their boogie boards to the left side of the punt and Dad would pull them along in their wetsuits with the outboard. She lost her grip and rolled under, blinking the water out to watch him turn around in a wide arc to pick her up. She had to wait alone in the open sea. The water was colder below the knee. She peered down her nose. Deep, dark blue. Scary.

She had been out with him on the *Samantha Rose*, when he put the ring net out catching pilchards. They were so pretty and silver, splashing around. Then from 13 onwards on the *Prevail*. She talked more about the sea than art and design. One day she told her tutor what she had been thinking for a long time.

'I'm leaving. I'm going to go fishing with my dad.'

'Well you go on about it enough,' he laughed.

Her college friends could not believe it.

'Oh my God. You're going fishing with your old man? That's crazy.'

'It's good money,' she shrugged. 'I'm getting bored here. If I get bored then I get moody.'

'We're going to miss you.'

Her friends from Cadgwith weren't surprised. They knew she was a bit nutty. She got her craziness from her dad.

Her alarm went off. She frowned at the caravan curtains; it was still dark. Three in the frigging morning! The night-time road was empty when they drove down to Newlyn in her dad's truck. The day before she'd grinned as she grabbed chicken, sausages, teabags, coffee off the shelves of the Co-op; enough food for three or four days. She nodded at the fishermen around Newlyn. Always fishermen around Newlyn. They bought some bait from the fish market then steamed out to wherever they needed to go. There's As and Bs and Cs in the sea; they were out in the Cs which was the furthest away. They did a bit of ring-netting for pilchards, but when they were difficult to catch they put the shark lines out, went into the Scillies.

She loved it out there. She loved the different seas. The way the rain skittered over the surface. How the water darkened and churned. The windless, glassy calm before sunset. Scud vapour followed them like gulls. Sometimes the water was so white and bright it made her sleepy. That white-light was part of seasickness, but she did not feel nauseous anymore; the trip back from Ireland had cured her.

She fished with her dad on the *Prevail* for eight months.

It was good money. The most she made was £300 on a three-day trip.

Her dad was great fun to be with. He'd tried to teach her knots: bowline, rolling hitch, whatever – they never stuck in her head. One day they went into St Mary's on the Scillies and he told her to tie the mooring line up to the bollard on the quay. She could never remember the knot, so she tied it like you tie a shoelace.

'What the hell is this?'

'I forgot.'

At the stern she tossed another mooring line to a boat on the quay, to pull them alongside. She waved at a man in the wheel-house. He waved back. She was standing in her oilers with a peaked cap on. Then she turned around and the man could see her blonde ponytail sticking out the back.

'My God, when you put your hand up I just thought it was a man,' he told them later in the pub. 'When you turned around I thought bloody hell it's a woman!'

The sharks followed the mackerel, so her job on the boat was to puncture a little hole in the mackerel, to bait it up ready to cast the shark lines off the back. She never gutted the fish. Another one of her jobs was making sure the ropes were paying out properly when they chucked the dan flag over. The dan flag was a marker for their line: a flag on a piece of bamboo, a float in the middle and a concrete weight at the bottom. The shark lines paid out for three miles and in between they'd have a fag or a cup of tea.

Sometimes he had to sleep.

'I'll leave you in charge for a couple of hours while I go below.'

He put on the autopilot and told her to check the plotter.

'If another dot appears, it means another boat is heading towards us. Work out where it's going, then steer away so we don't hit it.'

She read her book, but kept an eye out for dots. She stayed on watch for five hours while he slept. She was so good he left her in charge again. This time they were south of the Scillies, in the shipping lanes. She wasn't too worried since he was only off brewing a tea. The bloody radio was really getting on her nerves. It was an old car radio and every time they steered left or right one of the wires fell out. They were too far out to get radio signals. She wasn't allowed to play music when they were catching pilchards: it scared them off. It had a tape deck. Before she'd found some tapes she played as a kid and stuck them on. Funky ones like Leftfield or Massive Attack. She was listening to Norah Jones and looking out at the sea. It chilled her out so much, she drifted off into her own little world. Then this shadow fell over her. She glanced round through the back door of the wheelhouse and nearly had a heart attack.

'Dad! Look! Oh my God!'

There was a flipping great oil tanker right on top of them. From the shore they only looked a centimetre long. Tiny. This one slid over the sun like an eclipse; its steel hull bore down on them like a granite cliff.

Martin came in. He swayed from side to side so as not to spill his cuppa. He squinted up at the ship, saying nothing.

'We're already past it,' he said after a while. Sure enough white light streamed back through the wheelhouse window; the *Prevail* rose up, bucked by the tanker's wake.

At night Toots had her own bunk. She always had to sleep on her belly and let her body flop. If she slept on her back she rolled around too much. She tucked her trousers into her socks. She had a light, but when she got into her bunk her dad called out: 'Alright?'

'Yeah!'

He flicked the switch up top in the wheelhouse. The light went out. It was really cosy in her bunk. She thought of the students swotting for their course. She had made the right decision.

Then other crew came along with them so she didn't really do that much work, just went for the fun of it, earned twenty quid and was pleased enough. She was helping her dad out, but loved spending time with him.

'Do you want to come fishing?'

'Yeah!'

If ever he needed to go somewhere like Brixham or Falmouth to buy a winch or ropes or gear for the boat she'd go with him. It was a trip out, they'd get some lunch, or a pasty, play the radio in the car.

It was hard in winter. She hauled in miles of shark lines; as the rope ground the winch, she gripped the empty hooks, squeezed them open with their mechanism like a large safety pin and clipped them onto the bin. The older hooks were really stiff. Her hands were smaller than men's so none of the gloves fitted her. She had to use both bare hands to clip them on. With 500 hooks, her fingers went purple and froze right to the bone; so they lost momentum; the wind was biting and the boat rolling. God, she was so tired. She wished she was at home in bed. When they hauled shark lines in the dark she had to watch her footing. It was scary. She always wore a life jacket. The line was two feet underwater with a concrete weight on it. Her dad steered the boat. When there was a shark there, he tapped his foot.

'Bloody hell, Toots!'

She leant over and grabbed it. It weighed a ton. She felt the strength drain out of her as it dragged down in the water. Sometimes she yanked it in. Other times the boat rolled so hard she couldn't grab it and it slid out of her fingers and went by.

'What the hell did you do that for?' he snapped, coming over to help.

'I don't bloody know. Christ's sake.'

They'd have to go back and grab it.

The biggest one was seven feet. She had to roll with the boat and lift it in with the gaff. She had a feeling their fins went to China for shark fin soup.

She remembers as a child waiting on the beach for his boat to come back from the sea.

'Oh, Dad's back!'

He'd tower over them, bearded, grinning, playing the clown with his yellow oilers. He'd sit on the beach with them. They'd go swimming, or have a picnic.

Back home from one fishing trip she rang her older sister, Morwenna.

'What are you up to?'

'Just out milking.'

'God. See you later.'

'Hey. What are the names of my horses?'

Morna always tested Toots about them. Toots smiled and shrugged.

'A couple are big, a couple are small. They come and go.'

Morna was more like their mum: sensible, a land girl, into horses, ended up with a farmer. She got seasick out there, hated it. Toots was like Dad: crazy, a sea girl.

They had a hundred cows on Gwavas Farm, near the church, at the back of Cadgwith. Toots used to help Morna make yoghurts: put the strawberry or toffee or whatever in. She had a sealing machine to seal the top of the yoghurt. It was fun, like a small factory. Then the milk-pasteuriser bloke left and Toots said, 'I'll cover for him.'

She had seen him do the work. He'd used a massive great machine to get the milk up to a certain temperature, then bring it down. Skimmed came out the bottom into one tank, then he piped the creamy milk into bottles. She started fine but one of the bottles got crushed in it and went round in a circle. URNNNHH. Then the temperature button on the machine didn't work, so she pressed it again and again and again and the temperature went too high. CHCHCHTT. It took ten minutes for the temperature to go back down. Toots started at 7 a.m. and was there until one in the morning. A nightmare. She hung her head in her hands and cried. She was 20 now. It was hard work riding up the hill to Gwavas Farm on her cousin's old-fashioned chopper bike, but fun on the way down.

From 17 to 22 she lived in that caravan beside her mum and dad's place. She ripped all the seats out to make it a bit bigger, comfier. They had loads of parties up there which was great. Their poor old mum's bedroom was just through the caravan wall.

'Turn the music down!' she said.

Toots went with her dad to the Swordfish before fishing or after fishing and sat there with Ben Gunn, Jackie, Joe Crow, Dee, loads of other different fishermen. They put the jukebox on, had a dance or sat outside. One night in the Swordfish with Stacey they all got minging. They kidnapped Ben Gunn for a couple of days, brought him to Cadgwith for her dad's 50th. So much beer, wine and smoke that Ben ended up with gout. They had great times up there, with her dad joining in. Up until 6 a.m. Singing, dancing, smoking, drinking. Her dad's driftwood shack filled with people. Things at the Cove Inn would finish at one or two o'clock and everybody would wobble up there and carry on. Typically Toots, her dad, his cousin Luke and pals. They had a game where you had to chuck beer cans underarm into the coal scuttle. Singing the old songs, mostly AC/DC, Meatloaf and other old-school stuff on the stereo. She and her dad standing up and mouthing a duet in front of everyone. 'Martian Boogie' by 1970s Michigan rock band Brownsville Station. '*We're going to boogie, like those Martians do.*' It was really funny.

One night her dad leapt up, twisted round as if startled by a noise.

'What the hell is that coming down?!' he said, feigning shock.

They looked up wild-eyed as a mass of feathers flew out of nowhere, whistled across the room.

'Oh my God – what is that!?'

Martin had let the phez-chick go. To make the phez-chick he had dried out these chicken feathers, put formaldehyde on them, stuck them onto a big water bottle, made pheasant wings coming out of it; his auntie had knitted a chicken face; it had a metal beak nose. It had a little hook on top, so it flew from one corner of the sitting room to the other on a piece of wire. It'd be fixed up on one corner, then he'd let it go. Everyone would be sitting down partying on, then out it came.

He also had a kind of stuffed fox: a piece of wood with a fox skin around it and a fox's head on top. The rest was wrapped in rabbit skin. There was a tube going through its nose, coming out of its bum. Her dad put a fag in the fox's mouth, then sat around the corner and puffed at the tube, so it looked like it was smoking. It's head moved on a wire. They had lots of lots of laughs up there. They all creased up. It was brilliant. When Toots got her breath she said to him: 'You're like one of my best mates.'

'Bloody hell! I thought I *was* your best mate.'

'Well, you're one of my best mates.'

The cove is typical Emmerdale Farm; people know people's business and all sorts of things go on. She moved away twice. Once a friend got her a housekeeper job to a billionaire in Texas. She'd never met such an arsehole in all her life. He was horrible. It lasted five days. She went to New Zealand for two months, round the South Island and then to the North Island, to stay with her third cousin from Cadgwith who was doing well as a chef there. But by autumn her work hours had dropped off, she'd pretty much run out of money and her sister and friends were posting photos of Kennack Sands on Facebook. It was sunny, flat calm. So she came home. She hasn't been away since, not even up Camborne where her boyfriend lives.

She walks down the hill in summer past the whitewashed thatched cottages and it's packed. Holidaymakers say to her you don't appreciate it. But she does. That's why she stays. She feels sorry for the emmets because they have to go home. She wonders which towns and cities they come from. She spent four days in London once. It was fun but how much sightseeing can you do? You needed money to live there.

In January all the boats were pulled up and all the shops were shut. It was still beautiful. All the sea and storms came right up the alleyway. You felt the full fury of the ocean. Mountainous waves amassed at the mouth of the cove, cantering in from headland to headland. As they hit the rocks they exploded like bags of flour. Wooden doors splintered, stone walls were hammered with battering rams. It felt like a fistful of gravel was being thrown in your face. It was wild, exhilarating. The Christmas lights were strung like coloured beads from the gables and scattered around the cove; a neon lighthouse blinking. It was cold and it had character. Father Christmas came down the hill and they gathered to sing carols. Toots and other nutters got all dressed up, gathered as many signatures as they could then ran yelling into the icy sea. Afterwards, she drank a pint through chattering teeth, before heading up to the farm to Morna's. Toots cooked, gave Morna some jobs to do and got into trouble.

No job could touch fishing with her dad; those days were her fondest memories.

She can't face waitressing anymore in the Housel Bay Hotel on the Lizard or the Top House Inn in Helston. Like her mates she has three jobs through the summer down the cove. She works in the gallery, then for ten hours a week she solders circuit boards for amplifiers that go on luxury yachts.

Her mum asked her to fix a little crabber made of pieces of stained glass soldered together. She gave her money for the glasswork, soldering iron and grinder. At her workbench Toots thought – can I make a boat like that? She cut out a square of glass for the wheelhouse and a shard shaped like a boat, ground and copper-foiled it. What could she put it on? On her bench was a dark beach pebble. She put foil around it and soldered the boat onto it. The pebble lifted the boat up on the brow of a dark, curved wave. She bent a bit of wire for the A-frame. She thought that needs a dan flag like for the sharking. So she put a piece of material on a bit of wire, then a bead for the ball float.

Galleries sell them for £15 and take half. She sells them privately at the same price because you can't undercut the gallery. One year she was looking after her goddaughter, didn't make any and lost momentum. Now she wants it as her career. Last year she had them in one gallery, this year in five – Cadgwith, Porthleven, Penzance, the Lizard, Marazion – and next year eight places. Next week she's going to Port Isaac and Padstow with her dad. She's sold nearly 200. She should really curb her drinking and do her tax return.

Her girlfriends have more money because they have better jobs. A couple of them are teachers, one's an event manager for big parties. They are all going to Bali and Indonesia together for the winter. Toots could work her arse off all summer and go away for six months but she loves the winter too. Put the fire on and snuggle up and watch the soaps. Just before Christmas it is pretty quiet, just doing boats and circuit boards. Then January it's the daffodils. She doesn't pick them, she packs them. The radio's on. It's three months' good money. Then back to glass boats, circuit boards and maybe in the shop. She gets down days when she wishes she could be somewhere else, but she's been somewhere else and she loves it here. She's not really very materialistic. She'd like to go clothes shopping, but isn't really bothered. Her mate Holly, who's a carer, is the same. They have just enough money to get by. Lisa has got four babies and lives on the estate. She's a full-time mum. Tasha works in the Mullion Cove Hotel, cleaning the pub and crab-picking for her dad in summer. Her dad and brother catch them and she picks them: cracks the legs, gets the meat out, then sells the meat to a café down the cove. Toots's old mates Brett and Nugget work on the wind farms in Scotland.

She was sad when her dad retired from fishing, but was so glad she had the chance to go fishing with him. She lost the photos of those trips to the Scillies for years. A friend gave them to her husband. But they turned up.

Her dad's had his hard times but they've partied through. She puts bad thoughts in a cardboard box and burns them.

5

Sally had started running a tiny gallery in a fisherman's loft in Cadgwith called the Crow's Nest. She ran it with an artist friend, Gilly, who sold her own paintings; Sally took a cut if they sold work by other artists; with her pension it was just enough to live off through the rest of the year.

Martin kept fishing for four years on the *Prevail*; he was doing better than he did with the *Penrose*, but his knees were starting to be painful. He was doing cable watches to keep him going. His fishing licence cost £40,000 and he still had to pay off the *Prevail* and the insurance costs. When he left fishing in 2004, he was 50. Many skippers his age had moved out already. Martin was £96,000 in debt. They sold the council house to pay off what he owed and moved up next to Sally's mother's bungalow. They put a caravan in a field and next to it Martin built a house out of driftwood that had washed ashore from a shipwreck. It took all summer. It took two years to sell the *Prevail*. The person who bought it wanted eighteen inches taken off it so he could fish up in Portsmouth.

Martin found it hard that he had worked all his life but didn't actually own anything. He didn't have a penny to his name. He walked down the hill to Cadgwith Cove Inn on his own. He looked after Sally's mother when Sally was at the gallery. Then one day he told her – that's it. They split up in 2009. Martin explained to the girls that they had had a wonderful marriage for twenty-five years. He told his mother that he couldn't have wished for a better wife. But the fishing life had put them under pressure. Martin went home at the end of each day to stay in his mother's council house where his brother André also lived. Sally stayed in the caravan. They are better friends now. She is happy.

In the day, in the driftwood house, Martin is surrounded by nets, shark hooks and canvasses. He has been painting his life as a fisherman. The wind rattles through in the winter. His dogs lie curled by the wood-burner. Since he stopped fishing he has slaughtered pigs, run a bar in a container and now he is a painter. And a minicab driver. The bonnet is painted with the white cross of St Piran, the Cornish flag. All along the body of the car with a thick paint brush in a wavy line he has daubed his mobile number. He is still a well-loved character amongst the trawlermen and in the pubs of Newlyn. They all have time for Nutty Noah. There's a general belief that he never got the credit he deserved for keeping the old tradition of ring-netting alive.

Fishing in Newlyn can seem ruthless; when one person fails another steps in and takes his place. The other fishermen had been carefully watching Martin experiment with ring nets. When the *Penrose* sank in the winter of 1999, they were free to copy his ring-netting techniques without losing face. One fisherman's name was Stefan Glinski. He was an inquisitive fisherman who took risks. He was a thickset guy, balding and with a tan that made him appear ruddy-cheeked and jowly; he wasn't scruffy, but wore fine, pastel-coloured shirts. His father, Witold, a former construction worker, had escaped from a Siberian gulag and walked 4,000 miles across the wastelands of the Gobi Desert and the snowy Himalayas. In 2002, Stefan started ring-netting with a crew mate in St Ives Bay, where he lives. They came a cropper, mended the gear up and thought: we won't do it like that next time. They came down to Newlyn and found the sardines were literally just outside the harbour. One night they caught twenty-odd tons. Stefan wanted to work more and more. He pushed to go out four nights a week until his crew mate dropped out.

Stefan had researched new fishing techniques so he wouldn't capsize like the *Penrose*. He wouldn't overload, stored the block with care on deck. In the Newlyn pubs they wondered why he had got Spanish hydraulics and gurdies; but he had done his homework. The fish are sensitive to light, so they turned the lights off and worked in the moonlight, with only the glimmer from the sonar as they paid the nets out. Gulls screamed overhead. They tightened the purse strings around the shoal, until up came the silver treasure from the deep. He finished landing at midnight and was home in time to take his daughters to school. He went out Sunday to Thursday bringing in the catches. He bought a bigger boat, over seventy tons and forty-four feet, *White Heather*, a converted North Sea prawn trawler, and stored the fish down below in slush ice. If Stefan could stand up he went out; in winter gales and savage storms. He kept fishing pilchards for nine years. At 56 he was landing 1,500 tons, a third of the total amount landed by Cornish fishermen. None of them were thrown back into the sea.

Nick was right about pilchards. He teamed up with Stefan. Stefan was more business-minded, like him: Stefan had his own filleting and freezing facility, lots of expensive kit, three regular crew members and a relationship with wholesalers. Nick successfully registered 'Cornish sardines' with their own quota. The tins he processed were decorated with Newlyn School paintings, like Walter Langley's: decorative, sentimental works depicting fishermen's wives reeling from the news of a drowning. Waitrose began to stock them in 2000. M&S followed three years later. By 2003 prices for Cornish sardines had risen from one pence a kilo to a pound a kilo. Mevagissey caught on; a traditional fishing village, it revived its fleet by putting money into ring-netting for pilchards. More and more ring-netters set out from Newlyn, driven to find the pilchard shoals. Nutty Noah wished that Nick had lent him the money to buy a bigger, better boat rather than just £200 to buy the Madrons' net. A boat large enough to have a fish hold below deck to store the pilchards in slush ice. In 2004, the year Nutty Noah left fishing, Nick teamed up with other processors and Cornish fishermen to agree how to catch, process and market Cornish sardines. Stefan borrowed from his family and friends to buy a new boat and paid them back in eighteen months. He invested in a small facility in Hayle where he filleted and froze the pilchards.

Stefan was a 'scaly back', as they are known locally, born in St Ives. St Ives had always been different. Fishermen from Mousehole and Newlyn bury their differences to mutter about St Ives.

'St Ives men swim like this,' Edwin Madron said in the Ship in Mousehole, clawing his fingers inwards. 'To gather up all the money. Money-grabbing, religious St Ives people. I don't go down there, it's too busy, nothing but one-arm bandits, bloody surfboards. Years ago you park on the front, see a couple of fishermen, have a chat, a cup of tea, then come home again. Now you can't even find nowhere to park. They've done it themselves with all these bloody ice creams, cafés and one-arm bandit machines. Down there it's such a big place, the youngsters are on drugs. They fall out and fight.'

St Ives shopkeepers say everyone is jealous, because St Ives is the most expensive resort in the UK and their season is longer than anyone else's.

Only a cobbled forecourt separates the fourteenth-century Sloop Inn from the sea. One of Cornwall's oldest pubs, it's been a fishermen's haunt since 1312, long before the artists' invasion 500 years later. It is low-beamed, with tankards along the bar, colour-washed granite rubble walls that shelter many ghosts. On his stool tucked away in the snug, every day at five o'clock, is a clubbable figure in his late 70s who knows all the secrets of these narrow streets: Harding Laity. There were loads of cats when he grew up. The old doors had circular holes in them for the cats to go into the fish lofts; the fish were stored with salt and the rats attacked. 'The fishery', as his grandfather used to call it, everything was for 'the fishery' – the train wasn't allowed to hoot as it came into St Ives in case it scared the pilchard shoals. Serious things, passed by Acts of Parliament.

Since medieval times St Ives had been the key fishing port on the north coast, with a fleet of 450 boats. There was a divide between Downalong, the fishing quarters, and Upalong, where the tradesmen, incomers and well-to-do lived. The man who had a hosiery shop exactly on the divide knew which direction the ladies came from just by listening to their voices. As a boy growing up Harding wasn't allowed down to the lanes of Downalong, because there were rough men in smocks with carts clattering over the cobbles. They went seining for pilchards, three boats at a time with crews of up to seventeen, hauling in millions of pilchards for export in hogshead barrels to Catholic Italy. Now there are fifty boats.

St Ives' position is unique. Going around Land's End puts a ship in tricky, dangerous waters. Many ships were smashed to splinters on the cliffs, the Atlantic edge. But St Ives was the only safe haven on the north coast; Padstow, much further up, had too many treacherous sandbars. So they came across from Ireland, down from Wales and even from Spain or across the Atlantic. It was a vast harbour and became very wealthy: a huge meeting point with a terrific mix of people. There was a lot of Irish blood, two of the biggest families were the Quicks and Stevens. For the postman it was a nightmare; nicknames were the only way round it. There was a huge amount of falling out, a lot of it around religion. Schisms kept being formed as people argued, until by the 1920s there were nineteen denominations there – enough for a large city. The talk was sophisticated, humorous and revolved around inns and pulpits. People had sharp brains from eating fish and there was a lot of good talk. Debate. Talk. That's what built St Ives' culture.

Brunel built a bridge across the Tamar and from 1877 people fled there on the Great Western Railway's branch line, away from the smog of industrialised cities. Victorian seasiders carried a picture in their mind of the quaint ideal of a fishing village; they'd seen it exhibited in the Royal Academy in London. They responded to the glorious prettiness of the harbour, the light, and beyond the town, out on the moors, the mythic power of an ancient landscape. They were escaping something.

Artist colonies began in the 1920s. They were running away from the rise of the Nazis, fascism, the shattering effect of the wars. Barbara Hepworth fled there, afraid of bombs falling on her children through the glass ceiling of her Hampstead studio. When the M5 was built in the 1960s, people escaped the cramped housing of Birmingham, Bristol and Manchester. In *To the Lighthouse* Virginia Woolf drew from her family holidays in St Ives, with a garden down to the sea and Godrevy lighthouse as her inspiration.

Their talisman was Alfred Wallis, a former seafarer with no formal training, who started painting when his wife died. His primitive, naïve pictures were a welcome break from metropolitan sophistication. All around them were desolate moors, rugged cliffs and the relentless, restless ocean.

Harding became an estate agent. He had the monopoly in St Ives. He saw it all. Over decades he sold off every granite fisherman's cottage, every whitewashed pilchard cellar and net loft in Downalong. Like in Cadgwith and Mousehole, the old Cornish fishing families moved from the labyrinth of cobbled streets that led to the quay, up to the top of the hill, to pebble-dashed council houses, 'homes fit for heroes'. It's a different community up there, the original St Ives community, now so old they can't make it down the hill, and shop at the local Spar by the coach car park. In their place came rich incomers from the Home Counties, stoked by massive City bonuses, lawyers on six-figure salaries or canny roofers from Manchester; they invested in holiday lets close to the shops, the Tate Gallery and beaches. The really upmarket St Ives people moved to Clodgy Point with its amazing views of the sunset, or Beach Road where they stepped out onto the vast field of sandy beach, Porthmeor, a magnet for surfers and families. They don't want to live in Downalong; they've risen out of that. One resident bought an end-of-terrace place on Godrevy Terrace, near Porthmeor Beach, divided up his front garden into four parking spaces and sold them for £50,000 each.

When Harding started out they put the electric in, scrubbed everything down because of the fishing, put linoleum on the floor, usually had a painting on the wall of a steamship with an H on the funnel for 'hunger'. Other 'pierhead' gouaches depicted Cornish ships under full sail, with a smoking Vesuvius volcano in the background, delivering barrels of pressed, salted pilchards to Italy. He can't value anything anymore because the tiny cottages he sold for £24,000 are now going for £300,000.

Most are looking for views, but not many have them. One local lady, Rozzi, had done up a lot of houses in her time. In Teetotal Street she divided one place into a flat down below called Duck Down and a maisonette above, Duck Up. She sold them both on a ninety-nine-year lease and retained the freehold. Normally the freehold rent is £100 a year, but she insisted instead on two magnums of champagne on 1 May in time for her birthday on the 16th. Someone wanted to give her four bottles once but she refused: 'No, no, no, no! Harrods despatch magnums anywhere in the UK for £15.'

So on her birthday everyone brings a bottle of fizz and goes to her party in Teetotal Street. She is going to leave the freehold to her nephew.

After forty years living full-time on the edge of Downalong, in Norway Square, no one is living near Harding now. When he walks out in February he bumps into men with little spots of paint on their spectacles.

'Having a lovely week?'

Harding addresses them in his booming, posh baritone, threaded with only a skein of his Cornish roots.

'Lovely. Done two rooms.'

Harding finds the incomers all very pleasant. The incomers tend to add rather than take away. They'll buy a boat. They'll be in town and lobby to keep St Ives on an even keel. Mr Pegotty's was a nightclub in St Ives where bands performed in the 1970s. For local young people who danced there it is one of their formative memories. To Harding it was a bane: all the youth would come to Downalong then go back up at 1.30 a.m., peeing in people's letterboxes and waking everyone up. Not very nice. He was pleased when a solicitor, a lovely chap who earned £700,000, bought it and turned it into three units. That took the blot out of Downalong. Harding rang the solicitor's office and they said he was in the States. Harding was impressed when he rang back after lunch and said:

'I'm looking out over Central Park.'

These people have the ability to put pressure on the police force. Stop St Ives turning into Newquay. There is one late-night bar in Fore Street which is notorious for waking people up but they might not get a licence renewed. Youngsters have shots, come out and there's mayhem. It's the Newquay influence. Particularly in high season. Too much money, too little education. *Youth!* Harding is turning 80.

Harding runs the five o'clock club in the Sloop Inn. It is hilarious. Colonel Mike, whose wife was from St Ives originally, all you'd expect from a retired colonel. Derek, the solicitor, from the middle of Devon. One's off to Amsterdam today, the other's flying to Venice tonight. Penwith Pete, from the Penwith Gallery, the key gallery in St Ives. A lovely chap, used to be a surveyor, has let his hair grow long. His latest exhibition is by a Cornish painter who was a fisherman. A whole gang. Harding knows all the history of the cobbled streets. He has been teasing the Tate for ages about their £10 million extension which is utter nonsense, just to make offices, all underground. Ridiculous.

The standards are very high now. Harding has never known it as busy as this year. He bumped into one family: a mother, father, two kids and a dog, who were renting a cottage for £1,500 a week. You could go to Wogga Wogga for that and have guaranteed sunshine. Half-term? Who could afford to go away at half-term? When he was at Hayle Grammar School he might have had half-term, but didn't go away. Harding keeps in touch with his school pals who are now in Porthleven and St Just.

They've all got two heads over there. Foreign parts. Rough as rats. After leaving grammar school many of them became teachers. One followed the other. St Ives has spawned dozens of good schoolteachers. It was a massive change from sons of fishermen following their fathers into fishing.

Harding has a friend, an ex-miner, who went to Africa and designed diamond tables, ones for sorting out the top carat from the rubbish. Interesting man. A quarter of a million Cornish skilled miners went abroad when the tin and copper mines declined.

Harding was a figurehead. He was asked to speak at a mayoral do some years back. He stood up and glanced at his notes. He had a good story about one of his Mousehole ancestors, the infamous Dolly Pentreath, the last fluent Cornish speaker. Stories about her insulting people in Cornish went down well. Dolly used a hatchet to fight off press gangs trying to coerce fishermen into joining the navy, and saved a fugitive from hanging by sheltering him in the chimney of her old stone house overlooking the quay. Harding was about to tell how she called the vicar of Paul 'old frogface'. Then he noticed that the current vicar, seated next to him, looked a bit froggy. God! He went on quickly, missed out that bit.

Harding's grandmother was the youngest of five children, in a one up one down in Mousehole. The doctor could not tell her mother this fifth child wasn't going to live because she couldn't feed her: there'd be a clamour. But doctors were more than doctors in those days; they were psychiatrists, everything. So he said: 'You've got an unmarried sister in St Ives. This child needs the fresh air of the north coast.'

There was of course no difference in the air in St Ives, but his grandmother was sent there and the childless aunt took her under her wing. She grew up happily. She married a grocer down the harbour who supplied all the boats and gave credit to the families while the fishermen were away, up the Celtic Sea, through the canal down the North Sea. Young Scottish girls followed the kipper boats around, then settled in St Ives, where they married locally, brought new blood into the town. The Cornish were all marrying their cousins and in every farmhouse between here and Land's End there was a chap who wasn't quite elevenpence three farthings, but could lift a horse, help his father on the farm.

Harding's family farmed Tregerthen Farm, five miles from St Ives. A single, well-brought up young woman moved into the nearby run-down farm, Tremedda. She was raised in Virginia

Woolf's house in St Ives and had shaken her wealthy Scottish parents by developing a burning interest in farming. She fell in love with the local cowherd, Maurice, from the tiny village of Zennor. Her parents packed her off to India for a year, but she came back and promptly married him. She had the Japanese potter, Shōji Hamada, and some arty types round for dinner: a massive crock of rabbits cooked on a furze fire in the middle of the room. The cowherd said nothing while they discussed exhibitions. Only when they ate did he tell them how he had chased the rabbit with his hunting dog, jumping fences, running across fields, killing it fast so the meat stayed tender. He made the butter himself. They were spellbound.

They had four beautiful daughters who went to Paris finishing schools but ended up marrying local lads too. The latest generation are still at Tremedda, making clotted-cream ice cream from Friesians. The sons are working in the entertainment side of St Ives.

Every five years Harding dresses in his morning suit and top hat and heads a procession with a fiddler and customs man in full medieval velvet costume. A crowd follows them through the streets, holding up traffic. They scale the hilltop to the Steeple, an obelisk built by a St Ives eccentric. Fishermen use it as a day marker. There he dances with the widow of a fisherman, while the fiddler plays and ten young girls in white dresses dance around, hand in hand, their hair thrown back. It's lovely. Very St Ives.

Harding can't help promoting the place when he's on the phone.

'Oh you want to come down. No women down here, that's the trouble, all the hairy fishermen just hang around,' he tells a girl in Yorkshire. 'What's the weather like up there? Cold? Oh, I had to take my sweater off today. I'll introduce you to Penwith Pete.'

He rattles with pills. Up the surgery are all the people he was at school with.

The St Ives festival is good. Bob Devereux has an outside thing in Norway Square garden, and if you've written a bit of poetry you can go and stamp it out, bend someone's ear. Bob runs it well. His normal speaking voice is marginally effeminate, but when he starts declaiming, his grey mane shakes, he hops around angrily like Rumpelstiltskin. Daniel, Harding's grandson, frightens him to death with his impression of Bob stamping out poems. Bob's star has risen with St Ives' fortunes. He started selling deckchairs in the harbour, then opened the Salthouse Gallery, taking risks on new artists, with no money. After thirty years he'd become a wealthy collector: paintings all over his walls, ships' models and 700 pipes he sold to Wills, the tobacconist in Bristol. His ten grandfather clocks, on every landing and half-landing, go off dong, dong, dong. Twice weekly he winds them up. Harding goes down and plays Monopoly there on Boxing Day until the small hours, in the big living room up above a massive refectory table. As Harding bought and sold properties around the Monopoly board he couldn't help doing an inventory of Bob's chairs, all mother of pearl, the carvers, thousand-dollar vases, the mineral collection and cassoni.

Harding has curtailed his lifestyle now to look after his wife, Dee. He can't leave her for long now. They've been married fifty-two years. Dee is very happy. She can just about walk. He taps her foot to tell her which one to use to go down each step. He's putting in an Acorn stairlift for £600. He's getting it from a friend. Done the survey.

They sold all their paintings about two years ago, after friends had a fire and lost the lot. Original John Parks, everything. Harding and Dee sold theirs through David Lay, the auctioneer in Penzance and took the money and had their fiftieth wedding anniversary.

It is lovely living here, out of the main thoroughfare.

In winter he always used to put on an extra sweater, a bobble hat and a pair of jeans and go outside and make boats. If someone had reversed into a dinghy, he picked it up and did wonders with fibreglass. Everyone has a bit of hardwood at home; they'll never use it but it's too good to throw away. He got all those. Friends would carry it down, run it and scarf it in. He bought a saw and had all the tools. He had good fun but last year was really wet and cold. So he's sold all his boats and stopped now.

Across at the Sloop Inn, Harding spies his great-niece Ellie, working as a waitress. Her manager, Leah Churchman, has been there eleven years; she was born in St Ives. She can tell from the long faces when it's everyone's last day. One punter missed his lift, took a taxi to the station just so he could spend another two hours there. Regular punters come down throughout the year. Charlie, 77, comes down twice a year from London for six weeks. The September crowd is more chatty, down for golf, coach trips; it makes their holiday if Leah remembers their drink. They order a meat dish and get a raffle ticket with prizes of flights to Australia.

The staff work fast around the unsmiling punters, distracted from the rain by seasidey paintings and a DVD on a flat screen showing misty landscapes, helicopter shots of castle ruins on small peninsulas, aerial shots of crowded sandy beaches.

'Buggy has to be folded,' Ellie says to a harried mum. It's not a pub for kids: food doesn't come fast enough for worn-out, wetsuited, red-cheeked children who've been crabbing on the beach with bucket and spade and ice creams, running squealing from the breakers at their heels.

Ellie and Leah are part of a tight team that go out together after work. The younger crowd start in the Sloop but move on later to bars like the Hub or Balcony; at midnight they pound up the postered stairs and walls, under a ceiling covered in LPs and beer mats, to the Attic, which has a late licence. Kerry and Ben, both 23, went to St Ives school together. Kerry spends winter in South Africa with her boyfriend. Ben can't afford to buy in St Ives, so lives in Carbis Bay with his parents, away from the madness. Everyone knows everyone.

The Attic is run by local lad Leigh Champion, who worked for twenty-five years in TGI Fridays in London; he taught Ben how to mix strawberry mojitos, Jägerbombs and other cocktails until it became muscle memory. Like an icicle chandelier, upended glassware hangs on a strip of yellow light. Ben's laptop glows over his face as he taps through Spotify playlists: 90s garage to get a middle-aged hen night up dancing. On Friday nights he starts at ten with Pink Floyd, leans across the bar and pours tequila into people's mouths for free, mixes caffeinated Jägerbombs, five for a tenner. After that everyone from 18 to 60 is dancing on the tables to Wolf Mother. One night a Neanderthal builder from Crawley groped a dancing woman. She fended him off and he broke her jaw.

In the winter the Attic has a young local following: Ellie, Leah, the staff from the Sloop and Ben's girlfriend and the Lifeboat staff. The Lifeboat is on the waterfront at street level, so when the rain hits, tourists huddle inside round the heaters and order chunky, home-made chips and lamb burgers with views of the bay. Ben's been a kitchen porter, knows how busy the chefs are. After the rush, they need time to think, slow down. All the young locals have three jobs through the season. He's worked painting and decorating for people from Manchester and Birmingham who have ten properties apiece in St Ives. His girlfriend is saving to go to uni in Falmouth; they will all travel up there to party. St Ives is a warning, he thinks. The rest of Cornwall will follow. Now amongst the fudge and pasty shops they have ten retail chains: vintage homeware in Cath Kidston, outdoor clothing in Joules, yachty stuff in Quba, £400 for a sailcloth jacket. They even have Superdry and Wetherspoons. The independent, local shops are going.

St Ives is gloomy in winter. The waves are too big to surf. There is pressure on them to travel. Ben stays. He likes to spearfish grey mullet. It's frowned upon by freedivers to wear scuba gear and spearfish as it is not a sporting way to pursue fish, so Ben uses a snorkel and dives deep down until his lungs might burst. A big bass is worth a lot of money. The easy way into St Ives is to skate downhill; they skateboard at Camborne skate park.

There's a big drinking culture in St Ives. Ben's crowd start on the beach, play frisbee, catch the sun. They start in the Sloop, then visit Ben's mates behind the bar in the Hub. The evening often ends up in the Three Ferrets, a scruffy drinkers' pub in Chapel Street, full of old fishermen with no teeth who are really into music; few tourists venture in, put off by the drab signage on the front, rum locals loitering outside. Inside it is one room, a pool table, cheap beer and cheese and biscuits on the bar. They call it 'the Ritz', as in Fer-ritz. Frank, from Hair by Frank, goes in there with his terrier. You don't go for a haircut at Hair by Frank after lunch if you like your ears. Some joker took a ferret in there, but Frank's terrier chased it and bit its head off. The ferret's owner threw a brick through Frank's window.

Something always happens in the Ferrets. Ben got chatting to one toothless old fisherman called Willy who'd been away but moved back. It turns out he'd played guitar with Paul Weller, had a drink with Johnny Cash. Someone had their face slashed in there.

Cornwall is like a Christmas stocking – all the nuts go to the bottom.

Harding remembers the Ferrets when it was a late-night dive with lock-ins. The landlord pulled down a piece of fishing net over the bar when it was time, and if you wanted a pint he raised the net to take the money, pushed the pint through. There were three constables in there drinking up; it was always full of policemen. They were from local families, knew everyone. There was one rumour that a circus came to St Ives and a local girl was raped. The local men told the police to keep away as they took the rapist and threw him down a mineshaft.

On St Ives seafront, Lewis, 23, scoops ice cream for a dad in a T-shirt of Stewie from *Family Guy*. Lewis's seventy-seven-hour week is killing him. He serves Callestick Farm ice cream until the evening, then breakfast at the Kettle & Wink pub near the taxi rank, and he works at the carvery there on Sunday. He'll earn £2,500 over the six-week season. Some earn enough to winter on tropical beaches in tents; like the fishermen who live cheaply in India while the Atlantic storms rage back home. At 16 Lewis was sent away from the family home in Reading to live with his uncle in St Ives. He lives in a converted loft over-looking the sea with his partner and their kid.

The time goes faster for Lewis when Lauren is serving ice cream with him. She runs an ice rink in Canary Wharf in winter for drunk commodity traders. She is thickset with trendy glasses and black pixie haircut. They have a good gossip.

'Seven Stars in Penzance is dodgy. It has a pole for drunk girls,' Lewis says.

'Every pub in Plymouth is like that,' Lauren says. Her lips twitch into a nervous half-grin. 'Get your tits out and you get a bottle of champagne.'

'Would I get one with mine?' Lewis asks, his eyelids hooded, cupping his manboobs.

'Maybe half a bottle.'

The *Times & Echo*, St Ives' local paper, says this year shopkeepers are complaining about drunken fights.

'If I see another "lads on tour" T-shirt I'm going to punch someone,' Lewis sighs. 'Scouse stag parties.'

Lewis' mate's mum told him stories of being a barmaid in the Swordfish in Newlyn during the mackerel boom. Bricks of coke came in watertight packages from the seabed. The fishermen loved it. Wild, wild stories. There were drugs, illegal booze smug-gled in. The wives didn't know. St Ives has its own mutterings about drugs. One fisherman always went out fishing when everyone else had come in; he was scruffy, never dressed up. Then one day he disappeared to live in a villa in Spain. Another man from St Ives was rumoured to be bringing the drugs up to the north coast, but the deal went wrong and he was pushed into the harbour at Newlyn.

In the 1990s a lot of fishermen ended up with addictions. Mark Pod came down Newlyn quay when he'd finished learning as a deckhand, and said: 'I'm a skipper now, I'm taking this boat to sea.' Two months later he was found dead in one of the toilets from a heroin overdose. There is still a box in Newlyn's toilets for needle disposal. One ring-netter's young son crewed at 16, made good money through the 1990s, up to £1,000 a week. In his late teens, in with the wrong crowd, he dabbled in heroin and became an addict, ended up in Dartmoor prison. Drugs became endemic. Newlyn was the second worst place for heroin deaths after Liverpool.

There was a skipper of a beam trawler, a Brixham boat, who shopped his entire crew to customs for being off their heads on heroin. Another crew member was found dead with a needle in his arm in the ship's head. In the most dramatic incident, the crew of one Brixham boat were so out of it on heroin that the boat steamed into harbour, smashed through the other boats and straight up the beach. It was a sign of how hard the fisherman's life had become, how difficult it was to scratch a living on the smaller boats after the beam trawlers had dragged up the seabed, after the Scottish industrial ships came and fucked up the mackerel fishery that employed thousands of Cornishmen.

Drug dealers like the fact that Cornwall is full of tiny, secluded coves, but they still have to negotiate its dangerous coastline. In

2013, one Dutch skipper put in a Mayday call nine miles south-west of the Bishop. His yacht was towed into Newlyn with its sails in tatters. When customs boarded, he climbed the mast to escape and fell to his death on the quay. A search revealed 200 kilos of cocaine on board with a street value of £20 million.

On the seafront dads balance boxes of pizza, for hungry, red-faced kids with wet hair, and a chorus of Happy Birthday rings out from a dayglo-pink ice-cream parlour. Below the Sloop, a cluster of teenagers stand round on the corner listening to a man in blue shorts and salmon hoody. He looks like Keith Chegwin but sounds Geordie: 'It's a broken world. Jamie Bulger was killed by 8-year-old boys. The Devil is at work. When he's sorted out the paedophiles what about the terrorists? I've contributed to the world's pain. I've been rude to my sister. I've damaged my brother . . .'

It's Vinny from the Christian beach mission, which holds beach activities in the day then talks in the evening. A policeman leans out of the window of his cruising Corsa to tell him to move on. Excited children jostle down steep, cobbled lanes beside white-washed walls, draped with toddler wetsuits. Granite.

The crowd streams along Fore Street. Amanda, in a boutique shop selling hand-knitted jumpers, stayed away from St Ives a long time. Newlyn, Penzance and Marazion slagged off St Ives, labelled it commercialised. But she's a recent convert. She was running a caravan park in Marazion on old mining land. The more she put in the more she got out of it: one Kent family of regulars have broken through from being guests into being friends. Now, after her first season here, she prefers the buzz of

St Ives. She was surprised to find a very strong community that sticks together. She likes talking to the customers. Marazion feels slow now. She was born in Penzance and finds it a bit dirty. It is two minutes from Marazion to Penzance, fifteen to St Just. It's claustrophobic sometimes.

'I wish I was a fisherman . . .' a song comes from inside the Union Inn. Steve Jones, local singer-songwriter, stares wild-eyed through a cowl of lank, straight hair that reaches his shoulders. A St Ives veteran, he knows what his audience wants. He sings of the sea, giants, kings, saints and Celts. It's Wednesday and everyone has live music. 'In the Summertime' comes out the Queens Hotel in Bedford Road. There's a female vocalist in the Kettle & Wink.

'I wish I was a pirate . . .' croons Steve.

In the nineteenth century St Ives lived from fish. The three fleets were moored in the bay: mackerel drifters and the fast-sailed, lugger-rigged herring boats which followed shoals to Scotland and the Irish Sea. St Ives had thrived for centuries on pilchards. There were dozens of heavily built seine boats, up to forty feet long, coated in black tar, anchored on boat plots on Porthminster beach. The crew chatted, smoked and brewed tea all day until the huer's cry went up. Then they went out and enclosed the shoal in a seine net fitted with large corks on top and lead weights at the bottom. The pilchard fleet sent over 6,000 barrels to Catholic Italy. Towards the end of the nineteenth century the giant silver shoals went away. No one knows why. One theory is that they were attracted by the red water that flowed out of the mines, down streams and rivers into the sea. With the mine closures, the red water stopped. Net lofts and pilchard cellars, blacksmiths' and boatbuilders' workshops, all lay empty.

Then the artists started to move in.

First Turner and the marine artist Henry Moore in the mid-1800s. Then others. Greater numbers came when the railway arrived in the late nineteenth century. Its arrival meant they could ship their paintings to annual exhibitions in London, sometimes in specially reserved train carriages. In Newlyn, Walter Langley and Stanhope Forbes started painting the plight of fishermen and their families in picturesque style. The cheap studio space and tuition made St Ives increasingly popular with aspiring young artists. Many of the artist colonies in the late Victorian era were middle class and respectable, well connected to London's art institutions. Their studios were opened to the public each March. During and after the Second World War a younger generation of artists were drawn to the cliffs near Land's End, their ideals shattered as war raged in Europe. This wave hoped to find peace amongst West Penwith's ancient landscape and surrounding sea.

The journey itself was a major achievement in those days. But St Ives was worth it; with its four sandy beaches and charming cobbled streets, it was like a holiday on the French Riviera. With the exception of London, nowhere in Britain had so many artists living and working as Cornwall. They settled among the cliffs: St Just, Lamorna, Sennen Cove, Zennor.

St Ives was riddled with boatbuilders, carpenters, weavers, woodworkers. Bernard Leach, the potter, aimed to revive the lost traditions and 'spirit' of English country slipware. He made practical, beautiful pots at the pottery he co-founded in 1920 with his Japanese friend, Hamada. The alchemists of this age were potters. Cornwall is a geological freak with every mineral in its earth. Potters collect clays, china clay, and the variety they found there was a dream. In St Ives they could begin again. They could build something new, like children on the beach.

Barbara Hepworth left Hampstead a few days before the war began. She arrived in St Ives with her 5-year-old triplets, the cook, nurse, and her husband, Ben Nicholson. She was so preoccupied in St Ives with running a nursery school and foraging for food, she drew only at night. They moved in 1943 to a large, shabby house in Carbis Bay which gave them both studios. The children went to school; the bustle of family life gave way to the sound of mallet and hammer as she produced some of her best work. Her previous lack of space had made her obsessed with larger sculptures, over ten feet high. In her early 40s when the war ended, she watched the horrific newsreels of Belsen, with their pictures of emaciated human forms. The images haunted her.

She liked picking up pebbles on the beach. Their graceful flowing lines. On her walks she felt rooted. The standing stones and stone circles reminded her of the Yorkshire she had seen from her dad's car window. She explored colour in the concavities of form. Then in 1951 Ben left. Her response to the family break-up was to immerse herself further in her work. At auction she bought Trewyn Studio, a magical place behind a twenty-feet wall with a yard and garden where she could work in open air and space. This tiny, wiry woman could be heard tap-tap-tapping away at some monumental block of stone or wood or metal towering over her like a giant's toy. Mediterranean light showered down onto her, behind a screen of treetops. St Ives' narrow streets became Lilliputian as tiny figures wrapped up her giant sculptures and heaved them onto trains to send off to exhibitions. One piece looked like a nut in a shell; another a child in the womb. A seventeen-ton trunk of Nigerian hardwood was shipped to her. It was sawn up and stored around town. Locals saw a six-metre high aluminium boat hull coming out of her studio – *Winged Figure*, off to Oxford Street. She tapped away in her studio. Always working. On a short holiday to the Scillies, she tripped and broke her femur. She hobbled around St Ives with a stick. She liked a drop of whisky. She fought cancer. Each night she took a sleeping pill. It took fifteen minutes to work. Her cigarette burned for ten. One night she mistimed it. She set fire to her house. She died in the flames.

A young Cornish painter called Peter Lanyon was an apprentice to Hepworth and later Nicholson. He lived in the red house just behind Harding; in the tessellated porch is the word 'Lanyon'. Peter was a ragamuffin, but was sharp, respected. Harding remembers sitting next to him in the barber's chair as the other men teased him about his painting. Heaven knows whether he might have been the big flag-waving Cornish painter, up there with the best, if he'd lived longer. He didn't like Barbara Hepworth much; she was an incomer.

Lanyon's parents knew all the locals. His mother would be with the Salvation Army blowing their instruments in a gale on the front. She raised money for the lifeboat, worked in the community. Lanyon grew up listening to Chopin: his father's favourite. The house was filled with music. Slowly the music lingered on a cadence. Then exploded. Fingers cantering over the keys. The boy was startled by its outbursts of speed. Chopin was true to how he felt. He watched his father develop photographs in his darkroom in the Red House. He watched him carve intricate details on a doll's house. When they acted plays, Peter had to mimic different voices. They made things – St Ives was full of crafts. He was very keen to know how things worked. In his wartime service in the desert, grounded with migraines, he built new aeroplanes out of broken bits, tested hydraulic pressures. He constructed a glider with a friend. He made his own frames, ground his own colours, carved, bent metal and used glass; there was always a tactile, physical element.

When he had children he was very warm, generous and funny. A charming dad. He was lively and biggish, nearly six feet, which stood out in St Ives where they were short (but not as short as the people of Camborne). He'd take the children off to Newlyn with a little outboard motor, dive down and disappear for hours underwater. The children, left alone in the rocking boat, scoured the wide sea, worried he wouldn't come back.

As Lanyon went down the town to go drinking he would bump into a wide range of people. First family friends from Upalong: his parents were middle-class intellectuals who knew the well-to-do. Then lower, in Downalong, he'd run into the miners he grew up with. He used to play matador, a dominoes game, in the St Ives pubs. They'd take the piss out of him for talking posh; it didn't make a difference. He lost his Cornish accent when he went away to school, but he was a good mimic and could put one on. Locals asked him blunt questions. Other painters didn't have those kinds of down-to-earth friendships; it was good for him.

Lanyon identified strongly with the tin miners of St Just. St Just's windswept coast, with its ruined tin mines and megalithic sites, was his Calvary. It was a big deal to get there from St Ives.

At first he walked for four hours along the coastal path, then later he went by bicycle, then motorbike and finally by car. If you stood on Carn Brea, above Camborne, you could not believe the horrific noise; the smoke, the stench, fumes. The stamp mills went on all night, pounding lumps of ore into powder. When he looked out and saw a cliff he knew there were miners working in it. It made him realise he saw the cliff differently from another painter who came from upcountry, who just saw a cliff. He wanted to remain true to the experience of these people.

The miners could hear the roar of the sea above them as they hacked closer to the surface with their pickaxes in search of tin. One mine was only forty feet from the seabed and the sea later leaked in, flooding the tunnels. The drill, called the Widow-maker, is now fixed to the wall of the Star Inn in St Just. Queen Victoria went down there on the incline shaft.

In 1919, in the Levant mine, a lift full of men thundered down into the shaft, killing thirty-one. He painted a huge canvas about it, *Lost Mine*. One side is quiet; the other is fast and twisted with more colour. A bit like the Chopin that filled his childhood. To him it was tragic how the same men he drank with were forced to dig down deeper under the sea. All to make a bit more money for the rich mine-owners from upcountry. Although his grandfather had been a mining director from Redruth, his father had strong socialist principles which he passed on to Peter.

The miners weren't making any money. So when Peter looked at the landscape that's what he saw: miners below, inside the cliffs, risking their lives.

When the artists opened up their studios to the public once a year, all hell broke loose. The locals told the painter off if he got

the jib wrong, or some other detail on the ship. They'd be ridi-
culed for that. The fishermen had more respect for painters who
painted fishermen, who went into their houses, like Stanhope
Forbes and Louis Grier. The fishermen thought abstract was just
a load of rubbish and told the painters so. Lanyon's oldest son,
Andrew, believes the St Ives artists went abstract because they
needed to get away from all this criticism. They cut loose and
got so extreme that the locals said 'I don't want anything to do
with it.'

Peter Lanyon and other artists wanted to create paintings and
form a loose group, but the trouble was the individuals were so
spiky; the falling-outs were terrific. A good example was the
sculptor Sven Berlin. Bare-chested in a salt-faded Breton cap,
sunglasses and Lenin goatee, he chipped away at a huge block
of stone outside the Tower House, where he lived. Berlin was
bohemian, drank too much and was immersed in local politics.
One night after a pub session, he dived out into the huge crashing
breakers to rescue another painter who was being tugged by
strong currents towards The Rock. He found him afloat, singing
loudly. The council built a load of toilets alongside his Tower.
In disgust, Sven left St Ives, in a horse-drawn caravan. He moved
to the New Forest with his long-suffering wife. There he wrote
a bitchy satire, *The Dark Monarch*, viciously sending up his artist
drinking pals. Lanyon was depicted as an effete figure with burly
fishermen mates who he would use to settle arguments by crushing
people's hands. There were so many libel threats the book was
withdrawn from publication.

Harding sold his copy for £75. He remembers seeing Sven
sculpting when he was a boy. 'Want to have a go?' he'd say.

Another character was Roger Hilton. In 1966 the Hiltons arrived in Botallack, a hamlet near St Just, and Rose, Roger's much younger wife, abandoned her art to bring up their young family. Roger banned Rose from picking up a paint brush. He did bad things to people in relationships. He was unbelievably rude to people. He started fights over who was the most abstract. He once tore his canvas off the gallery wall and sledged down the front steps on it. His paintings were fresh, childlike abstracts of boats, beaches, horses and carts, nudes. He became drawn into ruinous drinking sessions with W. S. Graham, the Scottish poet; they used to work on people together like a pair of terriers, find their weak points and destroy them. They were artists at it; it was their hobby. It was very cruel. Drink left Roger palsied, bedridden in a clifftop cottage in Botallack. A bottle of whisky a day, 500 cigarettes a week. He pressed a bell. One ring for his son to come and change his paint water. Two for the other son to play chess. Three for his wife to bring him dinner.

Born in a tenement in Greenock, near Glasgow, W. S. Graham did not find fame and recognition but knew periods of great poverty in his Cornish cottage with no telephone, no car and an outside toilet. He had fans. He wrote about writing. He invented his own space. He wrote to his wife upstairs, letters to his dead friends and had a magical, word-drunk Dylan Thomas style.

During the war many artists left St Ives to fight. Peter Lanyon was among them. He came back after five years in the desert and found Nicholson and Hepworth had decreed a new rule: that all artists joining their elite group had to define themselves as either 'abstract' or 'figurative'. He was very angry that this was happening on his doorstep. It was a kind of fascism in art – if you go to the left you won't go to the gas chamber, if you go to the right you will. Why can't any artists evolve in any way they want? Lanyon was in between, using abstraction as one brush, as one tool. He wanted people to be able to find their own way, which might be a bit of this and a bit of that. His friend Bryan, the butcher's son, was born in St Ives. He had a brain defect so he couldn't see perspective. His paintings were unspoilt. Lanyon was fed up with all these artists trying to paint in a 'primitive' style. Their sophistication was ruining St Ives. Bryan was the real thing: he never looked at other paintings, never read art books. He couldn't see things any other way. Lanyon used to walk the coastal path and when he came to Ben Nicholson's house he would piss against it, hoping it would fall down. He hated how the incomers dominated St Ives' artistic scene and how they bickered over who was abstract or not. He broke away and formed a separate group of artists in Newlyn, called St Peter's Loft.

Lanyon was proud to be Cornish and would always take sides against the English. His friend Dr Frank Turk, a biologist and Renaissance man who had been to China to study Chinese seals, said there were three cultures in the world worthy of an encyclopaedia each: China, India and Cornwall. They were unique because of their ancient traditions. Built over 6,000 years ago, the hedges of Zennor are walls of stones with hedgebanks on top where plants grow. They are the oldest things on the planet that have been in continuous use. In the beads of the Phonecians you can detect which bit of Cornwall the tin comes from.

Teaching at St Peter's Loft, Lanyon rode in on a motorbike once and drove it around inside the room: just to make the point that he wasn't a fixed position. He wanted to be a racing driver, not a painter. His son Andrew remembers how he taught the same to his kids, taking the thread of conversation with them all over the place. It was annoying, but always fun. It kept them restless, not fitting in to fixed frames of thinking.

Lanyon spent more time with his kids than many dads, in winter and summer, making go-carts, chasing them, going off on different trips. Andrew remembers once near Botallack, when they were up high, his father ran down the narrow neck of the cliff. They watched in horror as he tumbled over the edge: he was always a risk-taker. Then moments later, his face peered over. Lanyon started going up in a glider so he could see the patchwork fields of Penwith. He took too many risks. In Somerset, on a training exercise, his glider crashed on landing and threw him out. He died of the injuries aged 46. He left five children: the eldest was 17, the youngest 10. It was extraordinary what he achieved in a short time.

Andrew learnt to have nothing to do with anybody else, clear off into the middle of nowhere near the Helston river, down the end of a dirt track. Don't join clubs. Keep away from art societies. That is a Cornish attitude. The Celtic frame of mind is sporadic development, farmsteads just out of bowshot of each other. That's their God, completely different from the English village green. They like to keep their distance.

On his first day visiting St Ives in the 1920s, Ben Nicholson wandered up a backstreet. There, through an open door, he discovered a 'very fierce, lonely little man of 70', painting at a table. Nicholson peered through the door. It was full of paintings on bits of paper and cardboard fixed with huge nails to the wall. They were like children's paintings: primitive, naïve pictures of ships and boats on the sea. The old man, Alfred Wallis, hadn't had any training. He'd been a mariner and had lived a life at sea. So when he painted a boat, upright, flying off the page, that's how he felt about the sea. He could go up to an artist who had painted the sea beautifully, look over and say: 'It isn't like that.'

Wallis went to sea as a boy of 9. The groundswell of the Atlantic must have been frightening as he clambered up the ratlines of schooners, the wooden slats strung between strong wires that held up the mast. At the top he would step out onto a single wire foot-rope under the yards to reef and unreef sails against the gales as the vessel made its way to Newfoundland. One storm was so fierce Wallis and the crew only survived by jettisoning all its cargo. The sloop and ketch, the light, wooden sailing vessels of the nineteenth century, were at the mercy of the elements. He painted them all: schooner, brigantine, barque and fully rigged ship.

At 20, in Penzance in 1876 he married a widow twenty-three years his senior, who had a brood of children aged 3 to 19. Wallis set up a marine rag-and-bones store on St Ives harbour front selling second-hand ship's gear and scrap metal. He was imprisoned for a month in Bodmin jail for receiving stolen goods when a steel ship, *Rosedale*, was wrecked on Porthminster Beach. When his wife died in 1922 Wallis started painting, for company. He painted the harbours, rocks and lights that he remembered as a seaman. He painted the pilchard and mackerel drivers who worked the Cornish coast, their drift nets shot at dusk. The foremast was lowered to stop the boat rolling; the mizzen kept her to wind. He used household paint on wooden boards and cards; gave the pictures to neighbours in exchange for food: they used them as kindling.

Nicholson quickly appropriated Wallis's boats-and-lighthouse motif. He bought them in bundles for a few shillings and introduced Wallis's work to Hepworth and other artists, including an assistant at the Tate Gallery. Wallis died in the workhouse in Madron in 1942. He was saved from a pauper's burial when an artist remembered Wallis had set aside money for a proper funeral. Ceramic tiles by Bernard Leach were set on his grave. Wallis's naïve images gave comfort to a generation devastated by war. They conjured up a lost, more innocent world. What people saw in Wallis's work was freedom. Freedom from training. Freedom of expression. There was the feeling of beach, space. Going to sea, fishing, was bloody dangerous, but you were free to do what you wanted. His work, born of the Cornish sea and earth, endured.

Nutty Noah and Ben Gunn both liked Alfred Wallis. He was a kindred spirit for both of them: a fisherman with no formal training who painted simple canvasses of life at sea. Martin arranged to meet Ben Gunn in the Swordfish to talk about art. He drove Ben over to his driftwood house to show him his paintings. The subjects were his life in fishing: the *Samantha Rose* in Cadgwith Cove, pioneering the ring nets in the *Penrose*, sharking in the *Prevail*. He carefully painted each white, silver fish in a vast shoal of pilchards. Ben left fishing because he had trouble with his back. Martin had now had his knees operated on.

Mackerel is a bloody hard living, all those little punts going out in freezing weather – rumour is that a gangmaster is running them now with Russians and Poles on £200 a week. Beam trawling is dangerous work and some of the boats are rust buckets. Martin doesn't like the stereotypes of fishermen as rough men in the pubs, pissed up the whole time, a bit thick. If you're out half your life on this bouncing platform, not eating properly, it's like a combat situation: you are uptight, living on the edge. Why wouldn't you have one night out on the piss when you land, and have a fry-up the next morning with everything? Most of the skippers from Martin's day have got out of fishing now. There are easier ways of making a living.

And fishermen have always had to supplement their income. Could Martin and Ben make a living out of painting? Can fishermen really make it as painters when they leave the sea?

Nigel Legge, part of the crab fleet in Cadgwith, who made withy pots by hand, managed to sell his paintings about crabbing. Like Wallis he painted on canvas, board, driftwood, and even tobacco tins. But Nigel had been the star of a TV series, *The Fisherman's Apprentice*. Nutty's own brother André made a living as an illustrator and did drawings for the packaging of Roskilly's Clotted Cream Fudge and the website for a campsite near Cadgwith. He lived in their mother's council house and used a room there as his studio. Their old man was a talented draughtsman and could sketch a ship well.

Ben painted on bits of card, canvas and even a kitchen door. He painted outside, on the gable ends of the Red Lion Inn, and painted the rocks blue by the lifeboat station. He met a few painters but couldn't mix with them at all: they were arty-farty and he was rough and ready. Ben had some luck. He had a local character to help him out: Shaun Stevenson. Shaun was Tony's son, Elizabeth's cousin. He always stayed away from anything to do with his family and fishing; he didn't want to know. Shaun was the black sheep, but he was well liked down the Newlyn pubs, in his torn jeans and his torn vest with a rearing unicorn tattooed on his arm. He was handsome with seared eyes and a lived-in face; he even had a fight with a fisherman over a woman.

Shaun had been very good to Ben. Ben found Shaun very fair. He heard Ben was starting painting. Shaun let Ben have a huge studio on the first floor of the slip in Newlyn. It was a massive place, a former pottery, that went back for miles and had a harbour view. It was like Picasso's studio. Shaun wouldn't take any rent off him. He had it for fourteen months for free. Then after that Shaun lent him another studio. Shaun helped him with an exhibition in Penzance and another one on St Michael's Mount. They piled all the paintings aboard a little boat with an outboard and off they went. There were a dozen in the restaurant in there. It's not a place for selling: people visit, see it, are gone. Ben would just have to keep on at it.

As fishing fell into decline, art was revived. Billy's daughter Claire showed Billy's private museum of dusty photos, anchors, pilchard presses and propellors to a painter, Henry Garfit, who was looking for new premises. He was stunned by the archive; it reminded him of the reclusive hoarder Howard Hughes. He loved the vast Victorian windows that let the light stream in: he wanted to start an art school in Newlyn. Henry had known about Newlyn from when he worked for a London dealer and wrote essays in their catalogue about the artists who migrated to Newlyn in the 1880s to paint the fishermen's working life at sea. It was a quaint, sentimental ideal of a fishing village aimed at visitors to the Royal Academy in London. Henry was so drawn to this art legacy, he resigned from the London dealer and started painting out of a net loft in Porthleven His studio was freezing and often awash with seawater. The calling was in Henry's blood: his father was a landscape artist who specialised in sporting rivers, while his mother painted ripening barley fields and elderflower bushes in Norfolk.

Today Cornwall's art history and concentration of galleries has drawn 10,000 creative people like Henry, working in unheated, damp places; a fraction of them making a living. They work for months towards a show. Most of them aren't paid, so their partners pick up the slack working as teachers and taxi drivers.

Henry approached the St Aubyn family, landed gentry who have owned St Michael's Mount and 5,000 acres of Cornwall from Lamorna to St Levan since the fourteenth century. The current family patriarch was using his City background to renovate old barns and outhouses for holiday lets or studios. For three years Henry lobbied the family to convert some barns into studios, only to see his request turned down for planning reasons.

Then he got chatting on the night train to London to the man in the bunk below. He was Alverne Bolitho, from a tin-mining dynasty, who lived a mile from Newlyn in Trewidden, a fifteen-acre family estate with gardens of well-tended camellias and rhododendrons open to the public. Within nine months, Henry was managing fifteen artists' studios, renovated from the Bolithos' stables and opened by Princess Anne in 2008.

With help from Billy's daughter Claire and the Bolithos, he turned Billy's private boat museum into the Newlyn Art School. Billy's archive of old boats, propeller shafts, mines, steering wheels and even the nameplate from Nutty Noah's sunken *Penrose* were cleared out as Henry made room for canvasses, palettes, easels and mannequins. Local kids from Newlyn's Gwavas Estate helped Henry paint the ceiling and dry-humped the mannequins. Henry aimed high when recruiting tutors and visited Naomi Frears, who worked out of a glamorous but decaying studio overlooking the sea in St Ives which once housed Patrick Heron and Ben Nicholson. Her work sold for thousands. But she agreed. It seemed even the best artists were pleased to have the extra income. Soon Henry built up a roster of Cornwall's top artists.

The punters signed up for the three-day courses: the school's most popular course naturally involved sketching on clifftops, hiking across the moors past grazing sheep. That tutor, Paul, an artist from Manchester, knew amazing hidden coves accessible only by crossing fields. They had great fun, munching sandwiches while painting ancient granite stone circles.

Next, Henry found an abandoned fish-processing warehouse. He explored the dark corridors, the light pouring into the wooden lofts and fish-smoking rooms. It would be perfect, he thought, for an ambitious exhibition of local artists, performers and poets.

For Ben Gunn the Newlyn Art School was expensive. It was £80 a day. *Eighty quid!* Way out of his league. Ben preferred a bench press to an easel. He had to move out of the massive studio Shaun had lent him; and found himself without a studio at all, which was a bit awkward. He had to move into the tiny fish cellar, a few feet square, in the basement of Ben and Jackie's Fore Street flat. He wedged all his dusty unsold seascapes in, piled up like logs. Stuffed them into beams. The stench from the drains punched up his nose. He did not get good vibes down there. The painting didn't go well. Christ, he could barely turn around.

Shaun tried to get him to exhibit at the PZ Gallery on Penzance Promenade. A mate of Shaun's ran it. It was a massive place, a vast 1930s art-deco building with high ceilings; used to be a car showroom. Ben could exhibit in there five high. Shaun's mate was a snob who felt nostalgic about the artists' colony of the 50s. He wasn't interested in Ben. But Shaun's other pal, editor of the *Cornishman*, published a feature about his painting.

Then one day Ben had some good news. The Newlyn Art Gallery was to hold an exhibition of local artists recommended by local people: Newlyn's greengrocers, the publican at the Tolcarne Inn, the harbour master, the cheese shop, and Patch Harvey from the lifeboat would all nominate their favourites. Then one of the locals chose Ben to appear. The whole exhibition was to be curated by the director of the Newlyn Art Gallery. It was some serious shit.

Ben reckons he was the only one who sold a painting in the exhibition. The buyer never got the canvas Ben showed him. He switched it with another similar seascape and sold the one he exhibited two days afterwards. Ben sold ninety paintings over the years. He looked at Nutty Noah's stuff. Nutty was very creative. He wrote rhyming poems about the history of pilchard fishing, how his great-grandfather, a huer, directed the seine boats from the cliff with gorse bushes. He painted his whole history as a Cornish fisherman, and his quest to find buried treasure on Kennack Sands. Nutty and Ben liked the idea that you needed to have been a seafarer to paint the sea, not to go to some poncy art school upcountry. Alongside his paintings, for good measure Nutty would sell CDs of him singing the old Cornish songs, DVDs, prints of his painting, nets, rigging and other goodies.

He was also a mole-catcher, chainsaw man, hedge trimmer and rabbit ridder.

Blair Todd, director of the Newlyn Art Gallery, is a slim forty-something, with whispery voice and cropped red hair greying up from the temples. He wore Converse boots, sandblasted jeans and a black long-sleeved T, with a white design so subtle it looked like he'd spilt milk down it. For him it was a difficult exhibition to curate. Fishermen rarely make it as painters.

He was brought up in St Ives in the 1970s, where it was not unusual to have a friend whose parent was an artist. At grammar school he was in the same class as Roger Hilton's and Bryan Wynter's sons. There were so many galleries now – around seventy in Penwith. Peter Lanyon's son Andrew joked there were so many artists in St Ives that it was impossible to paint a picture without other painters appearing in it.

Blair Todd does not see a strong artistic community in Newlyn any more. The majority of the artists in Newlyn are amateur, weekend painters. St Ives is more respectable, with more money than Newlyn. It is so gloriously pretty. People still go to St Ives to be part of an artistic community. They have made their money upcountry and move down to retire and be an artist painting traditional land and seascapes. Twenty years ago you could buy an old Methodist school for £70,000 then sell it eight years later for £700,000. House prices in the last ten years have become almost the same as in London. It used to be a ghost town in January and February, but now there's pretty much a year-round tourist trade.

Blair moved back down from London twenty years ago when there was no alternative cinema, no performance art, no experimental music, no cappuccino, no pizza. God forbid there wasn't even a kebab shop. In the same way that Upper Street in Islington didn't have a restaurant on it and now there's nothing but, Penzance has a tapas bar and arts club, and it's starting to spread out. Newlyn has a thriving cheese and charcuterie shop; if you said that five years ago you'd have wet yourself laughing. The Tolcarne Inn has been taken over by chef Ben Tunnicliffe, selling exquisite food. Even Mousehole has a deli now. It is very nice to cycle along the prom, get your cheese and cycle back. They are turning one of the old fish markets in the Coombe into an art-house cinema in partnership with the Curzon. Everywhere is becoming more like Islington.

At the Newlyn Art Gallery Blair is used to exhibiting artists with international appeal. Now with email and internet you can keep in contact with your dealers all round the world. Relatively big names like Ged Quinn and Alastair Mackie are based in Cornwall. Quinn, a Liverpudlian, takes a seventeenth-century landscape by Claude Lorrain and puts in a spacecraft from Stanley Kubrick's *2001: A Space Odyssey*. His mash-ups have given him an international following in Art Basel, a top inter-national art show, and the most cherished accolade of all, a glowing review by Michael Bracewell in the contemporary art magazine, *frieze*. Mackie is Cornish but his work has a wider following. His meticulously composed globe of mouse skulls extracted from owl pellets had exhibited at the White Cube in London.

Another artist who exhibited at the Newlyn Gallery was Jonty Lees. Blair found him fantastically charming and slightly bonkers. Jonty is in his early 40s, from outside Cornwall, but came to Falmouth Uni to do an MA and stayed. He wears a beret and has extraordinary facial hair. He lives in a National Trust cottage out in Zennor. In the upper gallery in Newlyn, Jonty did a three-day project where he built a Victorian velodrome and put a BMX in there. Then he planned to build a hot-air balloon using an industrial sewing machine, cutting out shapes and stitching them together. He invited everyone to join the Penzance Ballooning Club and get a badge. But he spent two weeks talking to everyone and barely stitched a thing. He draws people in with incredible charisma and childlike joy.

St Just attracts arty, creative types because it has always been much cheaper than anywhere else. It's right on the cliffs; it has always attracted the bohemians, and the New Age hippies.

Blair was born out on the cliffs in Botallack. In the winter the fog comes in and sits there for three months and is bleak and depressing, but in the summer it is gorgeous. When he was a kid St Just was the backward place where they had only just invented the wheel, hadn't got a telephone. The thick village. You had the hard-working miners, they seemed amazingly tolerant about the drop-outs, the people who found their way down there but couldn't work out how to get back out again. It was so far away from the rest of civilisation that they could stay in very cheap, tiny cottages. Twenty years ago, it had lots of New Age people walking around with the hardcore born-and-bred St Just. There are still dream-catchers on people's windows and horrendous things like that, but there are serious artists there for cheap property, beautiful sunsets.

In the last ten years St Just has attracted more people from Falmouth and Penzance who don't want to fanny around with buses and streetlights, who want to live the walking-on-the-clifftops life. (Although about five years ago a wave of artists moved out of Falmouth into St Just but realised that it is a bit bloody bleak and moved back into town.) Also, artists from London move to Cornwall, decide St Ives is too expensive, too clichéd, find there isn't much happening in Newlyn, don't want to live in a town like Penzance, so St Just is the natural place. It's fine living on a cliff, at least there is a Co-op for milk there, a Chinese takeaway, a few pubs off the square. You could still buy your Methodist chapel for a reasonable amount. There's a couple from the East End, a painter and sculptor, who teach at the Ruskin and Goldsmiths, who bought half a chapel, use it as a gallery for blocks of the summer, then go back and teach in London in the winter.

Blair loves Cornish artists who are rooted in London or in constant contact with international dealers. He wants artists who are reviewed in *frieze*. His taste is a million miles away from the landslide of seascapes that almost bury Ben Gunn in his stinky Newlyn fish cellar or the memories of pilchard fishing that crowd the low ceiling of Nutty Noah's driftwood house on the Lizard.

6

In St Just, many generations of mining families lived in tiny workmen's cottages up Carn Bosavern. Jeremy Le Grice, a painter, lived in Chapel Street, right off the central market square, in a three-storey house with a bay window, a studio at the back and a garden. Jeremy rarely mentioned his background. He was part of the Le Grice family who had lived for seven generations in the Queen Anne manor house, Trereife, tucked away at the top of the hill overlooking Newlyn and Mount's Bay, well out of the stinky fish market. Trereife had a cobbled courtyard, terraced lawns, tall yew hedges and 150-year-old wisteria. It was steeped in Cornish history. Jeremy had been pushed around Newlyn quay in his pram. When he was 3 his father died at Dunkirk which led to his mother having a nervous breakdown. His father's death haunted his later paintings, which were filled with dark silhouettes, black as Cornish granite. As the son of an Old Etonian who had been killed in the war, he was educated free at Eton. While he was at the Slade School of Art, radical 1960s politics – the student protests against the Vietnam War, the sit-ins against racism and banning the H-bomb – challenged the elitism of Eton so much that he hated the idea the locals in St Just might know that he was a 'blueblood'. But his family went back generations. How could they not know?

The Le Grice connection to the area had begun in the spring of 1796, when a new tutor arrived at Trereife house where the wealthy widow Mrs Nicholls lived with her sickly son. The tutor was a young Cambridge graduate named Charles Valentine Le Grice, who had abandoned his circle of glamorous London friends, which included the poets Wordsworth and Coleridge and the essayist Charles Lamb. They were horrified that he had given up his prospects for a wild place like Cornwall, and a dismal tutoring position. He was described as a jocund, rubicund little man, full of puns and jokes, very genial. Le Grice married the wealthy widower, Mary, in 1799 and became a curate. They had a son together. Le Grice outlived both Mary and her sickly boy, and his own son inherited Treveife.

The estate was eventually eroded as each generation faced crippling death duties and running costs. It passed to Jeremy's uncle, who employed local people to pick daffodils in winter in fields out towards St Just. He shipped truckloads of flowers out to Holland and was a well-known grower who became high sheriff of Cornwall. He rented the fields and sold his forty-year-old herd to pay off the farm overdraft. Le Grices at Treveife found the term 'blueblood' unwelcome, since their ancestor was a penniless Suffolk vicar.

Jeremy was strongly influenced by Peter Lanyon when he studied painting under him at St Peter's Loft in St Ives. Peter became his mentor and friend; they had a sensual time together with two Swedish art students. They both shared a passion for St Just and knew all the valleys and the sensational coastline. When Peter died at the age of 46, Jeremy lost his champion. He was terribly upset after his death. He was no longer introduced to all the dealers who came to see Peter.

It wasn't just the quality of the light that drew artists to places like St Just. It was the pre-Christian Celtic landscape, ancient stones. It was the cliffs, blue raging Atlantic swells, the lichen-covered granite. Denys Val Baker, editor of the *Cornish Review*, said it was more like a hidden, primordial force. Walter de la Mare said he did not feel safe until he crossed the Tamar back to Devon. Another novelist who lived at Sennen Cove, Ruth Manning-Sanders, said that the drowned sailors of the past could be heard hailing their names above the moaning of the waters. The sea is never at peace, it seethes against the rocks. Baker claimed there is anger there, that the Cornish are not part of twentieth-century civilisation: they belong to an immense, primitive Celtic past.

St Just was a lawless frontier town. The community policed itself. There was a story that a heroin dealer moved in and soon some of the young people were hooked. In the dead of night the local men came for him. They carried him, hooded, kicking and yelling to the edge of a mineshaft. They tied a rope around his ankle then threw him down. In one version of the story they pulled the rope back up and there was nothing on the end of it. In another they pulled him up and the dealer was so terrified that he left town. It was easy to lose your grip on the belt of a man who is bucking and writhing.

St Just is so remote it has retained the old Cornish traditions of rural justice. The locals used to call this 'rough music'. Any breach of the local moral code led to a noisy, unruly ruckus which invariably drove the offender from the community. Adultery and incest were punished in this way. In 1880, six men from Stoke Climsland were prosecuted for meting out this sort of justice, but were let off with small fines when their lawyer pleaded ancient custom.

A few years ago the biker chapter that ran St Austell took a man who owed them money out on Bodmin Moor, buried him up to his neck and left him overnight. Remote Cornwall can still be just as violent and lawless as it was in the days of *Jamaica Inn*. Local boys at school all know about being 'dangled' down the mineshaft. It's a rite of passage. Sometimes the reason for the dangling was a just cause. Sometimes not.

A traveller community pitched up near the edge of the cliffs with a view out to the Scillies. A local midwife, Rachel, went out to deliver a baby in the full force of a fierce gale. The travellers linked arms in a human chain on the cliff edge and passed Rachel along the line. She entered one of their bender tents made out of flexible willow or hazel sticks woven into a dome, with a tarpaulin over the top: a simple shelter, which stood up to the winds coming across the sea; a wood-burner provided heat. She said delivering a baby in that tiny tent was a marvellous experience.

Then local men went down and set fire to their tents. One St Juster couldn't find his getaway car in the dark, was caught and beaten up.

'The travellers were driven out by violence,' said the landlord of the Star Inn, Johnny McFadden. 'But they brought violence with them.'

The miners were a powerful force, a law unto themselves who were wary of outsiders. Their defiance was steeped in history. The Cornish economy depended on mining for centuries, fuelling the Cornish spirit of independence. Polytechnics were built on its profits; brilliant engineers competed hotly to perfect the best high-pressure engines for pumping, pulverising and crushing ore; drills were exported to mines all over the world. Richard Trevithick from Camborne invented the self-propelled steam locomotive. The German adventurer Rudolf Raspe wrote *Baron Munchausen* while he worked at Dolcoath mine. The mining devastated the landscape: red ground covered it like a wound. At the Crown Mines, the deepest shaft goes out 250 fathoms below sea level. The donkey pulled the winch and the miners sang as they went underground. One miner, Mickey, used so much dynamite that no one would work with him because he would 'blow the bollocks out of the whole place'. He was as deaf as a post.

The miners were superstitious, living with the dread of sudden accidents. A teenager in Tincroft mine fell forty feet to his death. A 72-year-old was forced to work down the mine because his three sons had been killed in accidents and his wife was bedridden needing care; a chain broke on a steam winder, sending a bucket down the mine with the chain still wound around the old man's arm. It wrenched his arm and shoulder off, exposing his heart and lungs. It was a hard, dangerous job. The horror they felt as the shaft lift rumbled down into the dark bowels was kept at bay with rousing singing, shoulder to shoulder. They sank into the gloom with only the glimmer of a taper to see by and a fierce Methodist belief in salvation to comfort them.

The tin miners were granted a Charter of Stannaries in 1201 which gave the miners their own government to safeguard their livelihood. Cornwall was granted legislative semi-independence. The miners were a powerful force, on the Cornish coat of arms. When Edward III created the Duchy of Cornwall in 1337 it confirmed Cornwall as an independent nation. Tin miners did not have to answer to English courts or the laws of Westminster. St Piran saw a white-hot molten cross bubble up when he put his hearthstone in the fire. He rediscovered tin-smelting. That white cross became the Cornish flag and standard.

When Jeremy Le Grice and his wife Mary moved back from London to St Just in the 1960s, he wanted to set up his own life, independent from his privileged background of Trereife, only six miles away. They were part of a wave of incoming artists that was more radical than the well-heeled ones who came to St Ives in the 1950s. For ten years they lived an idyllic life, two paint-spattered artists in their studios, with a young family. There were wild, boozy parties, with jazz playing and notorious bohemian figures like Roger Hilton. The Hiltons became close friends.

One of their best-loved haunts was the Gurnard's Head hotel near Zennor. The landlord, Jimmy, a decorated officer in the 1st Airbourne Division bought the lease with his army pay-off. On Saturdays he kept illegal hours through until dawn, only served customers he liked and had theatrical rows with his wife. His favourite fishermen drank all night on credit, crashed out in his bed, all on the promise of a bag of fish in the future. There was no cash register, just rolls of notes stuffed into cracks around the old Aga in the kitchen at the back. Jimmy hated men in advertising or TV who drove Jaguars. In winter, with driving rain, thick fog and a howling wind, the poet W. S. Graham staged impromptu one-act plays. Then the lights would go out. Jimmy would swear and fumble under the bar for old oil lamps. There was a potent mix of miners, fishermen, artists and farmers. They all drank long and hard. It had a wonderful atmosphere.

A powerful, larger-than-life figure amongst them was Herbie Uren. He was a stocky, ruddy-cheeked fisherman and miner, who also bred pigeons. If the mood took him he might climb on a bus with his mates and disappear for a couple of days.

Jeremy was not a regular in the pubs. He had a stammer. He went walking every day with his dogs in Kenidjack Valley, one of the finest industrial landscapes in Cornwall, with engine houses perched on the edge of dramatic cliffs, Iron Age cliff castles and Bronze Age cairns. He was intrigued by these ancient fields above Sennen, how they hid something mystical. He found walking there unique as the paths went by wild ruined mines. Thousands of shafts survived; thirteen engine houses were saved. He loved to go up Chapel Carn Brea, which lies just south of St Just and overlooks Sennen and the Atlantic ocean. It's 200 metres high and can be seen from the surrounding countryside and cliffs. He painted it as a menacing, domineering black peak rising out of his painting.

The sites were safe but in a ruinous condition. A walker had to be careful. Jeremy knew those paths intimately and could go out walking when the light was fading. Every night Jeremy walked out on the cliffs with his bull terrier. He heard the distinctive 'cheeow' of red-billed choughs. Peregrine falcons and Cornish choughs were breeding. His favourite part was the top of Kenidjack cliff with magnificent views along the coast to Cape Cornwall and, on a fine day, across to the Isles of Scilly.

Jeremy walked from the Count House just off the South-West Coast Path at Botallack. In the mothy dark he vaulted over stiles, plunged down steep descents. He passed a sign which read: 'Mine shaft danger of death'. In the fading light, landmarks were like ancient warnings: a ring of Bronze Age stones, the towering silhouette of a tall calciner chimney.

In 1969 his marriage broke down. Mary ran off with Herbie Uren. Jeremy was devastated. It took him a long time to recover. There was a feeling too in St Just that Herbie had done well for himself. Herbie was a huge figure down Geevor mine. He was also a Cornish fisherman: a 'real man' in the Lawrentian vein. He had charisma. The mining community was close-knit because they risked their lives going down digging for copper beneath the sea. It's the same risk in fishing. When Geevor mine closed, locals put bags of vegetables outside the miners' doors because they knew they were too proud to ask for help. St Just is so remote, so desolate, that people forge strong friendships to make it through hard winters. If a gale blew someone's roof off, their friends would climb up and hammer the slates back on. The local miners, exploited for many decades by rich landowners, slapped Herbie on the back; he had scored one over the blueblood Le Grice family by stealing off with Jeremy's wife.

For Jeremy, usually such a gregarious character, the atmosphere in St Just was very difficult. The break-up was very distressing. Damage was done. He was worried he would lose his children to Mary. He was living in Chapel Street off the main square, where some St Just locals closed ranks behind Herbie. Jeremy had some allies: the Boscean potters and his cleaner. He was still deeply embarrassed about his privileged background. All his life he constantly pushed back against it, and his radical art-school side felt living in the classless former-mining community of St Just was part of that rebellion.

At a New Year's Eve party that year he met Lyn. They had known each other before. Her marriage had also ended. She was teaching at Bournemouth Art School but her vocation was interior design. As the party moved toward midnight, they both suddenly realised their futures might be together.

She moved into the three-storey house in Chapel Street where he'd lived with Mary. Lyn loved the beautiful bay window, the garden. Her nature was well-suited to supporting everyone in the aftermath of the break-up. Many days would start with a visit from Herbie's ex-wife, distraught. She took some comfort having coffee with Jeremy and Lyn. She was a tough, primitive woman but she was raw with her emotions. She was desolate, having being abandoned with her two boys. She lived in one of the cottages, up Carn Bosavern. There was nothing there but a table and a chair: bare, without comfort. It was pure Lawrence, Lyn thought.

There were still parties at people's houses and they would go afterwards to the Gurnard's Head.

'Le Grice, out! Your bloody aunt's a magistrate,' Jimmy the landlord yelled at them once.

Jeremy felt he could not stay in St Just any longer. It had lost its flavour at that time. He was stammering terribly but with Lyn it went away. Lyn gave him a stability he had not had before. She was expecting their child. One day they set off to Falmouth to pick up a new car. Driving back past Mylor dockyard, Jeremy stopped. There out the window was a beautiful masted Baltic trader, a ketch called the *Heddy*: a sailing craft with two masts, a large mainmast forward and a smaller mizzen behind. It looked like one of the picturesque paintings of mackerel drifters by one of the Newlyn School artists. It offered return passage paid to Los Angeles for the crew. Was he going to leave Lyn behind, expecting their child, looking after his children from his first marriage? But Lyn knew he would come back to her.

He set sail from Falmouth across the Atlantic with six strangers and his cousin Charlie. They went straight into a Bay of Biscay gale. Squalls struck at night. They had to snap into action. Jeremy hauled down the mainsail quickly and used the mizzen to steady her. The old wooden vessel, built in 1904, sprung a number of leaks. After twelve days they made Madeira and caulked the hull. They picked up north-east trade winds sailing goose-winged for 3,000 miles to Grenada, then up the Mexico coast to L.A. When they made port he phoned her excitedly and challenged her: 'Come and meet me in Los Angeles.'

She asked her mother's advice: 'You should go,' she said. So Lyn left their baby Jude with her and flew to meet him in Los Angeles.

When they returned, they stayed with Lyn's mother in the Cotswolds. Jeremy liked the villages in yellow Cotswold stone, nestled in the hills and old beech woodlands.

They found a beautiful set of medieval barns and began to renovate them into a home with enough space for all the children. They kept Chapel Street on while they finished the renovation.

Lyn, raised in a family of seven, loved driving her gang of six children from her and Jeremy's previous marriages off to Sennen with a huge great hamper singing all the way. Cousins and neighbours would latch on. Every midsummer, on Cape Sports Day, the cove was invaded by fishermen dressed as Vikings, their horned helmets silhouetted against the fog of flour bombs pelted by wetsuited kids. Thirty swimmers, from age 10 to 60, were ferried a mile offshore from Cape Cornwall, out to the Brisons, the two protruding black granite peaks of a submerged reef. They jumped from the boats and raced back to shore: a hell of a swim for kids against tricky currents. Only those over 14 were supposed to attempt it. Jeremy swam slowly, chaperoning his daughter, Harriet, only for her to pull away from him and churn towards the beach.

On Sundays they had the loan of Jeremy's uncle's yacht. After deck cook-outs or picnics all the kids slept below. They swam in rock pools. Jeremy's son Tom, who built his own Mirror dinghy, sailed around St Michael's Mount. It was a Swallows and Amazons childhood. Tom became such an accomplished mariner he ended up designing destroyers. Jeremy drove Tom and his two sisters to Trythall school, an old-fashioned, small family primary school with one teacher. The ultimate dare was bombing down the sheer Nancherrow Hill on a bike. The kids had terrible crashes, broke their collar bones.

Then they moved. Jeremy taught at Cheltenham and Hereford. It wasn't long before he felt restless. St Just had been pure Cornwall to him. In Newlyn he had painted the Stevensons' rusting old hulls, the salty stink of the fish market, the ebb and flow of the tides, the seaweed-draped ladders down to the boats at low tide. The quay was the centre of his shambolic life. But the Cotswolds left him landlocked and trapped. It was all so tame. He hated what he called 'the tight-arsed school of merchandise production'. He barely painted for fourteen years.

Mary had also given up painting. She had married Herbie and moved to Porthleven where she brought up their family for ten years. In 1972 Herbie bought the *Ibis*, a traditional Cornish fishing lugger, forty-two-feet, with splendid pitch-pine planking on oak frames: she'd been built in 1930 by the legendary boatbuilder Percy Mitchell, who was known as an artist in wood. Each plank ran the full length of the hull and rang like a tuning fork when struck with a hammer. The seam was so tight there were no butts to leak or caulking to fall out. Built for pilchard drifting and longlining, she had a suit of tanned sails, rigged Cornish Dipping Lug style, like the ones in paintings by Walter Langley. Herbie worked the tricky waters of the Lizard peninsula crabbing in the summer and mackereling in the winter. People always approached him when he moored the *Ibis* at a new harbour. She had touched many lives in her history, and had distinguished owners and fishing records to her name. Sailing out of Porthleven harbour he faced directly to the south-west open to the full force of the strongest Atlantic gales. He had no sounder or radio, just a compass and the wheel.

She was sound and seaworthy in poor weather, but Herbie was a boat-killer, not a maintainer, and after a few years he laid her up. The wheelhouse was now rough and empty of gear. In the engine room the fine eighty-two hp Gardner diesel was worn down, the batteries flat, the chain steering ceased and it was full of junk and rubbish. In 1978, Herbie sold the *Ibis* back to Paul Greenwood, her original owner, for only £2,500. Paul gave her a good overhaul for two weeks: repainted her, built a new wheelhouse, fabricated a trawling gantry and lead blocks. He then earned a living off her for twenty-three years and she became one of the last working luggers in Looe harbour.

Herbie started up Lifeboat Day in Porthleven, where kids dressed as pirates with collection boxes for the RNLI. Winch cables from Culdrose air base were stretched taut across Porthleven harbour. Herbie then rode across them on a customised post-office bike.

Lyn was driving near St Buryan with her mother when they came across some barns in a smallholding. She called Jeremy.

'I've found you a Cornish farmhouse.'

'Was I looking for one?'

It was the intervention of their son, Jude, that clinched it. He'd been born in Cornwall. 'I've always wanted to be a proper Cornishman,' Jude added. So they went back.

Lyn worked hard with her original stencil designs. She was happy with her big family. She socialised with the artistic community and people who came to stay. Jude successfully bridged the gap. He knew all the locals. They were so central that everyone constantly dropped in. Jeremy was close to the painter Karl Weschke. His first wife became a great animator. Their son would carry on St Just's creative legacy.

Jude dropped out of school at 15. From his local friends he adopted a certain air of Cornish defiance. He was a tall, imposing giant with curly hair, and spoke slowly in a strong bass. He became an apprentice to a local woodsman. With its windswept moors and exposed uplands where black granite tears through the surface, Cornwall is too desolate for many trees. On the Lizard peninsula, the green and red serpentine rock forms infertile soil. But in one sheltered valley on the

Trelowarren estate there is an ancient woodland. The Vyvyan family had owned Trelowarren for 600 years and were friends of Jude's parents. Jude loved the outdoors and the exertion of being a labourer. He found the woodland magical: above him soared Douglas firs, sweet chestnuts and the Mediterranean oaks planted in the nineteenth century; sunlight streamed through the canopy of branches. He looked down over the rolling white mists that swallowed sailing boats on the Helston river. He crept into a mysterious fogou, a gloomy underground chamber with a single entry, beneath an Iron Age hill fort.

Every year St Just has a two-day feast, a 700-year-old tradition and an excuse for a drunken punch-up with Pendeen. Everyone comes home for it: to the Cornish it is more important than Christmas. There are Methodists who only drink on feast day. One feast night, as singing and laughter echoed through the Star Inn, Herbie ambled up to Jude in the square and held out his hand. In his palm he cradled a model of a rowing boat.

'Pilot gig,' Herbie said. 'They were early lifeboats.'

He told Jude how in the seventeenth century men rowed out in desperate seas to carry out heroic rescues. The six-oared gigs started in the Scillies carrying pilots to ships wanting to negotiate difficult Cornish waters. Then in Bristol and Manchester, crews raced out to be the first gig to get their pilot on board and be paid. Jude was intrigued.

'Head down Cape Cornwall,' Herbie urged him, prodding his ribs. 'They're selecting the gig rowing crew.'

Jude went down to Sennen Cove. He was broad-shouldered and sturdy from clearing forests. Next to him stood hefty farmers, fishermen and other big local lads. The crew were selected on their strength, fitness and commitment. They clambered aboard and heaved through the grey waves until they were red in the face. It was a speedy craft. When they launched their first gig in 1990, Jude was on the team. His crew mates were a lively bunch of likeable characters. It was a special time for him; he glimpsed a real vitality across the county. Newquay had a particular buzz about it. You could roll up anywhere and bump into people you knew. Girls were involved too. Jude felt very fortunate to be amongst his rowing pals and felt a strong sense of camaraderie. When they rowed against Cadgwith, there was a fantastic atmosphere afterwards in the Cove Inn. He bellowed out sea shanties, hymns and songs from the tin mine in St Just in his wonderful, clear bass. He had an exceptional singing voice and had been in choirs since his teens.

For the best part of the 1990s he rowed gigs in the summer, then played rugby with the same guys in winter. Even in tough conditions they rowed right out, cut through the groundswell that reared up and bucked Jude so hard the oar behind jabbed his spine. Cold, grey water sloshed over them. Gigs overturned in the surf if the cox wasn't paying attention. It was exciting.

The sport expanded rapidly into regattas throughout the southwest and world championships on the Scillies. But, for Jude, it didn't have the same feel about it later in the decade. Strong, opionated local characters tended to fall out. The guys were still around, but their lives had lost that sense of good times. It wasn't just that the fire of youth was dimmed. The 1990s were bleak, terrible times to work the land. Two debt-ridden farmer friends killed themselves.

His parents arranged for him to have lessons with a professional opera singer from Cornwall who loved Jude's voice but was frustrated by how long it took him to learn the songs. Jude's learning difficulties would prevent him becoming a professional singer, he said. Jude became interested in the origin of music like Cherubini's Requiem, and decided to travel to Italy to search for better luck. He drove to Milan in a Land Rover, then down to Rome. On the outskirts of the city, he was alarmed to see a toll booth ahead. He'd run out of money. What was he going to do? He veered up a steep bank and smashed the gear box. Crawling on with a secondary set of low-speed gears, he cut through a fence with wire-cutters and rolled into Rome on a side road. That night he slept in the Land Rover and sold it for parts to a team of geologists. His Cornish friends put him in touch with a Welsh soprano, Anna Risi, who was astounded by his singing voice, which she found magical. She put him up. He draped his Cornish flag outside his window and woke to a view of the Colosseum. He attended church services and developed a deep religious devotion and reverence for the institution of marriage. It was under threat: Anna's circle was decadent in their relationships. Both his father's and mother's first marriages had failed. He needed to revive some of the old traditions of courtship.

Back in England, his opera tutor sent him to a week-long masterclass in Canterbury. He sat on a chair against the wall and watched a 23-year-old tomboyish drama student, Rebecca, sing the role of Cherubino, the lovesick adolescent page boy in Mozart's *Marriage of Figaro*. Jude was bewitched by her voice. Rebecca noticed this giant figure in a lumberjack shirt and huge boots, looking like a Thomas Hardy character. He turned up on her train home to Dorset, but sat staring out the window, jaw clenched with emotion, unable to speak. She agreed to give him her address, but felt so uncomfortable that she left the train a station early, and called her father.

Jude resolved to court Rebecca in an old-fashioned way, to show her Trereife and the Cornish landscapes his father painted. He rang her to meet up. She wasn't interested. She was curt.

He was staying with family friends nearby in Dorset, and asked his hostess for a ring. She didn't have one, so he took a taxi into Dorchester. He found a jeweller, kicked in the window, stole a ring and left behind a scribbled IOU on the back of a business card. Covered in blood, he set off to find Rebecca's house. Her parents were away and Rebecca had called friends and met them for a drink at the Stour Inn, in Blandford. By chance, Jude, tired, lost, laden with luggage, wandered down the road and into the pub. On seeing Rebecca he stood up and gave her the ring. Alarmed to find him there, she shared a quick drink with him, made her excuses and fled.

Jude was convicted of criminal damage and paid a fine. He then went back to Italy and stayed once more with Anna in Rome, where she taught him arias. Over the next four years, between trips to Italy, Jude was to visit Rebecca's house fifteen times. Rebecca's mother came to dread hearing his knock on the door.

'Rebecca is not here,' she told him. 'She doesn't want to speak to you.' He persisted. Once Rebecca did speak to him but she found his lack of empathy frightening. In Italy, Jude bought an old Mercedes truck, a Unimog, which had huge tractor tyres for tackling rough terrain. It was like an amoured vehicle that could carry huge loads, while hitting speeds of nearly 60 mph on the open road. Like Don Quixote, Jude adhered stubbornly to a medieval chivalric code in the face of an immoral modern world. Chivalry disposed men to heroic actions and grandiose gestures, not casual flings that cheapened the institution of marriage.

In 2001 he decided to make a more epic gesture. In Wales he used a crane to mount a thirty-feet ash tree onto the Unimog's roof, then drove a hundred miles to her house. The tree was a gift for her: a symbol of fertility and of marriage. He arrived in the small hours, unloaded the tree onto the lawn in front of Rebecca's house and began to dig a hole. Tired from his labours, he lay down on the grass next to the hole and slept. Rebecca's stepsister found him and called the police.

Six months later he left a Jaguar car in Rebecca's driveway at night. Jaguar was a car with a tradition and bloodline, a noble courtship gift, even though he'd bought this one for a few hundred quid on Penzance quay. He brushed aside warnings about the Harassment Act, a recent piece of legislation. The west Cornish had always defied new laws.

In 2002, Anna Risi took him on tour with her to Egypt. He sang the bass solo in Rossini's *Petite messe solennelle* to packed houses in Alexandria and Cairo. It was the highlight of his musical career. He returned in buoyant mood.

He saw that a local church in Blandford St Mary's had a display of vintage wedding dresses. Jude, smartly dressed and with one of the dresses tucked under his arm, tore off from the church in the direction of Rebecca's house. He felt it would be a funny gift, a comic touch after the previous failed gestures. They didn't see it that way. Instead he found he had to undergo a series of psychiatric assessments and sign on every day at the local police station.

In 2004, he set sail from Newlyn on the crabber *Julian Paul*, bound for France. It was one of the Harveys' fleet, an old wooden crabber with beautiful lines. It was skippered by Barney Knight-Trembath, one of Jude's rugby pals from St Just. He disappeared abroad for six weeks. The owners would not even have known he was on board. One dark night the *Julian Paul* put to sea from the Brittany port of Le Diben for the return journey to Newlyn. The Cornish crew had all been drinking and the local Bretons could hear them singing on board as the boat left harbour. The crew did not expect this giant of a man to have such an expressive, intimate voice. Barney was nervous. Le Diben was a horrible cove to navigate, so narrow, with huge rock formations sticking up both sides like fangs. There was nothing to guide English boats in or out, just a couple of day markers on shore. Bretons, who knew the waters well, steered themselves in by shining spotlights onto white marks painted on the rocks. This was far too primitive for most English skippers, who turned away. Barney had heard that two years ago the same *Julian Paul* hit the cliffs, and the skipper had to row ashore. Now as Barney bent down to flick on the plug for autopilot, he felt her crunch into a rock. She started to flood below. The crew had seconds to cut the life raft free before she sank. They were lucky. When they arrived back in England on a ferry, Jude was taken away. He disappeared.

Some time later, a couple called the Hudsons were hosting a New Year's Eve party out on the cliffs of West Penwith. The guests were becoming happily sozzled. At midnight a towering figure suddenly appeared at the door, flanked by two other burly men. He strode into the living room and faced the guests.

Then he began to sing in a magnetic bass. The guests were spellbound. No sooner had he stopped than he would launch into another song. He carried on from midnight until five in the morning, barely pausing for breath. Who was this splendid chap, they asked, with the voice of an angel? Then just as suddenly as he had appeared he vanished with his two men as dawn was breaking through the mist over the moor. It was Jude. The men, his minders, escorted him back to where he came from.

At a dinner party Jude's parents heard that Rebecca had married. Given his belief in the sanctity of marriage, they decided to call him with the news. He heard it like a death, a relief. His pursuit was over.

He hasn't seen her in eleven years.

In 2012, Jeremy died. Jude sang at his funeral.

In the final matches of the Six Nations rugby, the front room of the Star Inn in St Just is bursting. In a side room watching Ireland, the stern-faced barman Johnny McFadden, in full Irish shirt, sits with two compatriots. Cash bets are tucked in the rafters. The ceiling is draped with the flags of the real six Celtic nations, including Cornwall and Brittany. A huge bearded figure sits with a floppy Guinness hat on. Among the ex-miners and fishermen, old school friends, his broad-chested height isn't noticed. Johnny McFadden eyes him carefully, muttering that if he wasn't a blueblood he'd be sectioned. Barney greets Jude at the door; he's been on an oil rig in Baku and urges Jude to head out there. Jude mutters about his movements being restricted.

'Jump on a crabber and head to Italy,' Barney grins. They share a knowing look. Jude has begun to feel better. He's moved forward. In 2014 he took part in the annual Wall of Music festival, a five-day event held in the Methodist chapel in Hayle. There were classes in woodwind, brass, strings and piano. With some encouragement Jude decided to enter the men's solo singing competition. He wore a pristine white dress shirt and neatly trimmed beard. His cheeks were tanned. He chose a Victorian song; his singing voice was captivating. He won the competition and his beaming face appeared in the local paper.

They held an exhibition for Jeremy's paintings at the Penwith Gallery in St Ives. Penwith Pete thought Jude was a good man. There was a reception and the art tourists pored over the pictures. Jude managed to buy back one of his father's paintings of the Kenidjack Valley, which they both loved dearly. He still walks the cliffpaths like his father did with his bull terrier, and can dodge the deserted mineshafts even in the dim light of dusk. From Kenidjack he can see the ruins of tin and copper mines: Botallack, Levant and Wheal Bal.

Sometimes he passes the church in Zennor. It has looked down on the sea since the sixth century; a tombstone in the tower is inscribed for a 'Hen-pecked husband' with a beautiful engraving for the 'four winds that daily toss this bubble'. A yarn is told that a beautiful woman in a long dress sat at the back of the church, listening to one of the choristers. One night she lured him down to Pendower Cove for a dip and their singing can still be heard in summer. On the end of one pew is a 1,000-year-old carving of a mermaid holding a looking glass in one hand and a comb in the other.

Stonemasons worked on the church in the thirteenth century and a pub was built for them nearby: the remote, windswept Tinners Arms. In 1916, D. H. Lawrence and his German wife, Frieda, stumbled off the cliffs into the inn, where the blazing fire danced over the low beams and dark wood; its terrace overlooks the sea. Lawrence wrote:

> At Zennor one sees infinite Atlantic, all peacock-mingled colours, and the gorse is sunshine itself. Zennor is a most beautiful place: a tiny granite village nestling under high shaggy moor-hills and a big sweep of lovely sea beyond, such a lovely sea, lovelier even than the Mediterranean . . . It is the best place I have been in, I think.

They rented a cottage in Zennor for £5 a year and filled it with second-hand furniture. He worked on *Women in Love*. Kate Mansfield stayed next door but hated it and left after a few weeks. Lawrence railed bitterly against the war he loathed, telling the locals the newspapers were full of lies. His wife warned him to be careful. In the 'Nightmare' chapter of *Kangaroo* he described how the Cornish started to suspect they were spies: a man with binoculars lay behind a drystone wall with a lass eavesdropping; at night men crouched below their window and listened. Inside, Lawrence defiantly roared out German folk songs. Frieda's cousin was Manfred von Richthofen, the notorious German flying ace, the Red Baron. As war paranoia spread across Cornwall, rumours about them intensified: the flickering cottage lights were sending signals to German submarines lying in wait; the smoke in their chimney, and their washing on the line, were coded messages;

they kept a secret stock of petrol for U-boats. The poison took hold. They were stopped and searched by a military patrol, their square loaf seized on as a camera. The police searched their cottage and ordered them to leave Cornwall.

Lawrence said that Cornwall 'isn't really England, nor Christendom. It has . . . that flicker of Celtic consciousness.' Cornwall gave him second sight. At night the blackness of the moors called softly to him, the druids of the past spoke from ancient, pre-Christian stones. Old myths support the idea that dark spirits were about. In the winter storms on the headland of Treryn Dinas, a strangely shaped stone, the Lady Logan, was said to rock to and fro, moaning with the unquiet spirit of a giantess who knocked her husband over the cliff, where he lay howling, torn to shreds. Lawrence saw lights from carts and bicycles weaving across the tops of hills before a Spanish coal vessel of 3,000 tons was wrecked on the rocks below their cottage and local farmers carried the coal away up the cliffs in sacks.

In the late 1950s, a couple of American tourists made a pilgrimage to Lawrence's cottage. It is a remote, desolate spot. Nearby is Tregerthen Farm where Harding Laity's uncle farmed. They knocked on the door. It was opened by a middle-aged man who bore a striking resemblance to Picasso. In a German accent he said his name was Karl.

'Is this the cottage where D. H. Lawrence wrote *Women in Love*?' one of the tourists asked, wide-eyed. Karl nodded.

'Are those the shelves that Lawrence referred to in his letters?'

Karl ushered them through to the kitchen. Through a window in the living-room wall, he watched as one American carefully unscrewed a cup hook to claim it as a souvenir.

'Look, I saw what you did,' Karl said, coming in. The tourist reddened, hastily replacing the hook. Unbeknowst to them, Karl had only recently moved in. He flipped open an unpacked box and dug out some mugs he'd bought at Woolies the week before for a tenner. He handed one to the visitor, who cradled it like it was the Holy Grail. Karl shrugged and allowed them to take the mug away.

Many years later Harding Laity bumped into Karl at a funeral. 'Was that story true about the cups?'

'Yes,' the painter replied. 'We were all bums in those days.'

Karl Weschke was a colourful character. He was one of the Le Grices' great friends. One of three illegitimate children, Karl was born in Taubenpreskeln, near the city of Gera. His father was a long-haired, gun-toting anarchist, constantly on the run, whom Karl met only once, aged 11. His mother, a barmaid, abandoned him to an orphanage at the age of 2, but was forced to reclaim him five years later. He then lived a brutish existence with her, pretending to be asleep in the narrow bed they shared as she entertained men from the bar she worked in. He slept on piles of dirty laundry, learnt to use his fists, and once ate roast cat, which he found delicious. This childhood left him vulnerable to grooming by the Hitler Youth. He rode on the running board of the Führer's limousine and fought in the war as a commando, but was wounded and invalided out. The Belgian orderlies who loaded him on a stretcher onto a hospital ship were tipping uniformed German officers into the sea. It was a British medical officer who saved him: 'For God's sake put him in some pyjamas,' the medic shouted.

Imprisoned in Britain's toughest prisoner-of-war camp on a desolate Scottish moor in Caithness, he was shown films of the recently discovered death camps. Moved from place to place he was finally sent to Radwinter, an open camp near Cambridge, where art became part of his rehabilitation. He designed theatre sets and went to lectures, where he learnt about German expressionists banned by the Nazis. He met the local MP, the flamboyant *Brideshead*-era Tom Driberg. He moved on to Wilton Park, a Churchill-inspired initiative to re-educate over 4,000 Germans and establish a democracy in post-war Germany. Driberg encouraged him to attend House of Commons debates. Determined to pay his way to stay in Britain, Karl became a gravestone carver, an auxiliary coastguard, and even a lion feeder's assistant in a circus. He visited the Royal Academy and the Tate, studied at St Martin's School of Art and decided to become a painter after travelling to Italy and Spain. From his inauspicious beginnings as a street orphan, he became well-read and cultured. In 1955 he moved to Zennor, to paint and sculpt.

Driberg had introduced him to a tank-driving sergeant major called De Vere. Karl was attracted to the soldier's daughter, Alison, who was an art-school graduate. She was modest, of slight build, with a bush of frizzy hair and tinted specs. She had an infectious giggle and a magical glint in her eye. Their friendship blossomed into marriage, and a son, Ben, was born in 1956.

His marriage to Alison did not last long, but they remained good friends. In 1960 Karl eventually settled into a small, isolated house on Cape Cornwall, overlooking the sea: the perfect backdrop for his paintings. He stayed there for forty years. The view from his studio was over the desolate moors with their crops of bracken or gorse, to the long rollers of the Atlantic. As a scuba diver who collected his own shellfish, he loved and respected the sea, and painted it in many moods using original earthy colours. On one dive, after a severe attack of the bends, he was given up for dead. Karl dismissed the attractions of the Cornish light and painted at night, under a sixty-watt bulb, with the curtains closed. He was a good friend of Jeremy and Mary, and later Lyn. Self-taught, he remained aloof from the St Ives school, which he regarded as 'Bloomsbury-on-Sea'. Karl's paintings betrayed his own hard, heartless life: fighting dogs, horsemen on the moor, scenes of rape, storms at sea, fire-eaters and drowned men. There was a threat of violence; barbarism was not far away. He evoked Penwith's past, cloaked in myth, soaked into the black granite. He did not use the bright tones of the St Ives group.

He had no strong commitment to the abstract, or art for art's sake; he sought out friends who were also outsiders: Roger Hilton, who was German-born, and the poet W. S. Graham, who lived nearby. He refused to revisit Germany until late in his life. After a documentary about Karl's life was broadcast on German television, a German woman got in touch with him. Her name was Rachel. Karl had not seen her for twelve years. She was his long-lost daughter, born in the 1950s. He never expected to see her again and had even named his youngest daughter Rachel in her memory. They were reunited.

He had five children. He brought two of them up single-handed. His oldest son Ben remembers that it was a hard living. They would have sandwiches, then he would sell a painting and there'd be steak and wine. Then it was back to sandwiches. Karl bought fresh fish from the trawlermen in Newlyn. Roger Nowell, a skipper and a real character, would come round and slap huge turbots down on the kitchen table. Karl cooked home-made latkes or potato pancakes and cheesecake. He marinated olives, bought salami from the best delis. There were always guests in that bare kitchen. Ben stared out at great swells of the Atlantic rolling in; the house was warmed by wood-burning stoves.

One day when Ben was 11 there was a knock at the door. A towering figure blocked out the whole doorway. Ben had never seen a fireman in full uniform before.

'Have you got a telephone?' the man asked. There were few around out on the cliffs. 'A tanker's gone down on the Seven Stones reef.'

He glanced at the boys from under his helmet. 'Want a ride down in the fire engine, boys?' They jumped up beside him. Ben's younger brother, Lucas, got to ring the bell. They drove over to Cape Cornwall and stood on the shore and waited. Slowly the brown sludge came in on the waves. It was one of the biggest oil slicks on British shores. The *Torrey Canyon* had snagged its hull on the Seven Stones reef. Ten days later the boys heard waves of RAF strike aircraft scream overhead as they bombed the slick to set it ablaze. The tanker broke in half and sank thirty feet below.

St Just did not have the sandy beaches or natural harbour of other resorts, but it had its people. The fishermen and miners put their lives in their friends' hands. Both Ben and Karl had to climb down sheer cliffs when they worked as auxiliary coastguards. They relied on local men to hold the rope. They were accepted because they stayed and contributed to the community. It was the same for everyone. If you were friends with the St Justers, you were firm friends. It was really bitter in the winter. The landscape was magnificent, but very hard to live in. You had to make your own fun. Ben's friends in Padstow thought of it as real Cornwall, because it kept its identity.

Alison, Ben's mother and Karl's first wife, forged her own way in animation. In the 1950s women were only expected to do the drudge work: ink in and paint men's animation. She kicked back and became Britain's first woman auteur and ultimately one of the best animators in the country. Her masterpiece *Black Dog* was made in Cornwall, inspired by the bleak landscape around St Just. It won her a large number of awards and international recognition. To produce a similar twenty-minute film, an animation studio would need about twelve assistants but Alison completed it using only one or two part-time helpers, working miles away from recording studios or camera labs. It is a surreal film that plays with the theme of identity. At the end, a giggling baby is about to fall into an abyss when magically, beneath its tiny waddling feet, a viaduct appears. The child dances and whirls with its mother, growing with each twirl. It is a touching scene. Ben knows it is about him. He worked as her assistant. When digital came she was stumped, being no good on computers, so Ben helped her out. He set up all the special equipment needed to convert her work to digital in the 1990s. At one stage he had it all in a caravan, out near Chapel Carn Brea. It must have been the remotest computer link in Britain. If the Cornish could produce tin and fish, Ben believed, they could produce digital content too. A Scandinavian company was trialling how to get broadband to peripheral regions and they had heard there was a cluster of creative people near St Just. With an antennae and a booster station mast in Sennen Cove, Ben helped pioneer broadband before BT.

On one project, Spider Eye Productions, a London company, asked to use Ben's computer equipment. Spider Eye was run by a couple, Erica and Morgan, who had their offices in Oxford Circus. They came down and worked with Ben in St Just while the project was ongoing. At first they stayed at the Swordfish in Newlyn, because Morgan's brother-in-law ran it. They loved that Newlyn was a proper working port with a fish market. There weren't that many left. They drank with Morgan's nephew, who was a six-days-a-week crabber.

'You don't want to go up to St Just,' they said in the Swordfish. 'They're all cave-dwellers up there.'

St Just was a proper town too. In the Star Inn there, they were warned off the Newlyn locals who they nicknamed 'buccas'. In Cornish folklore Newlyn fishermen left fish by the water to placate the evil spirit Bucca who created violent storms. Erica and Morgan loved the place. They were inspired by Alison's example, making a beautiful masterpiece like *Black Dog*, down in desolate St Just, away from the industry. Now they were doing a BBC series, so Ben sent the machines from Penzance to London for them. He worked with them, commuting to London on the sleeper from Penzance, returning at the weekend.

When it came to the second series, Spider Eye asked Ben: 'Can you do more work down there, in Cornwall?'

'Sure. We can find some local people.'

Erica and Morgan moved down to work in St Just. As Erica commuted back to London she looked forward to seeing people, eating exotic things she couldn't find in Cornwall. But once she arrived in London, she hated the sheer number of people. She dodged through the throng at Paddington and was jammed into someone's armpit on the Tube. On the streets, nobody looked at anybody, only down at the pavement. Dirt. Crime. She couldn't wait to come back to the cliffs again and see the vastness of the Atlantic Ocean.

'Why don't we move down here?' Morgan said to her one day after lunch on the beach. With broadband fibre optic in place, they may as well be here as anywhere. Animation files could be sent up and down the line. They only had to go back to London to find experienced voice-over actors. Soho was becoming too expensive; other companies were moving out. Their rent in Oxford Circus was £60,000 a year. They enquired how much it would be to rent the spacious top floor of the Old Town Hall in St Just. It was only £1,000 a year.

So they moved, bringing their empire with them. They took over buildings in St Just, including the old police station, and a one-time brothel for miners. With fifty staff and forty other freelancers, they became the biggest employer in St Just after Warrens bakery. They brought their clients and crew with them. There were so many roles involved in producing a whole TV series – sound designers, editors, script writers – but everyone was there, in one building. There's no way they could do that in the tiny space they rented in Oxford Circus. Erica and Morgan then moved into a house in Botallack, near the mines that go under the sea. It was a huge change for Morgan, who grew up bang in the middle of Covent Garden. They drive half a mile to work, with the sea out the window. They go fishing down there.

When Spider Eye started producing animation out of St Just, Ben found they had a lot of support from the locals. One of them was Jude's friend Barney Knight-Trembath, the fisherman who had skippered the *Julian Paul*. Growing up in St Just, many of Ben's friends were miners and fishermen like Barney. Barney was right behind Spider Eye. 'St Justers can do anything,' he said.

The fact that they were there doing digital production in St Just was further evidence that he was right. They had pioneered the mines in Geevor. Brunel had come there with Trevithick and installed steam engines to excavate the ore from under the sea. Men from Botallack went to help Brunel build a tunnel under the Thames, at Rotherhithe. They worked on the Channel Tunnel. A thousand men from St Just went to South Australia to mine – Ben had visited there on a singing trip with a local choir and had driven past a sign saying: 'You are now entering Little Cornwall.'

It was exciting for locals to be involved with serious animators. Spider Eye made their own TV show for Disney. They'd released a mainstream movie for Universal. Some of the young people of St Just were in a rut, so Spider Eye also took on a lot of local Cornish school-leavers. Some of them went on to careers in animation. The best one was Barney's son, Rikki. Barney told Rikki to go along to a barbecue Spider Eye was having. So Rikki approached them and came to them from school. At first he would do simple stuff like scan in a pencil drawing, colour it in and then put it together with a background. Morgan could see Rikki was very good technically.

'Go and do a degree in CGI at Bournemouth,' he told him.

Graduates who studied computer-generated imagery at Bournemouth had gone on to be world-class animators, even working in Los Angeles on big-budget sci-fi films like *Interstellar*. So he went there. While Rikki was at Bournemouth, Morgan continued to supply him with work and advice. Rikki would never have pursued it if Spider Eye hadn't been in St Just. They brought him back when they were making a TV series for Disney Junior. *Jungle Junction* was for pre-school-age children about fun-loving animals and vehicles: Zooter the pig scooter, Ellyvan, and Beetlebugs. Rikki loved bringing the characters to life. He'd loved watching *Tom and Jerry* and *Postman Pat* as a kid. They had been his inspiration. Rikki was so skilled that he became Spider Eye's secret weapon. The family vibe in the office encouraged him to do his best. In the end he was so good at it he could work anywhere in the world. He got a job working for George Lucas's company, LucasArts, which made *Star Wars*. Then he went to work in Singapore for LucasArts and was steadily promoted until he became a senior animator. He became more successful than Spider Eye. Barney was right – St Justers could do anything.

Spider Eye trained up a lot of other Cornish school leavers who went on to do degree courses. It carried on the creative tradition. Ben had a dream that one day Cornwall would be the centre for animation in Britain. The world had once been focused on Penwith's mining expertise, then its artists' colonies. Now, with broadband, there would be animators' colonies out on the cliffs.

People constantly wrote to Spider Eye, asking for work. When they scaled up to do a whole TV series and they needed fifteen animators, Erica picked the best ones then told them: 'You have to move here for eighteen months.' They jumped at the idea, and rented locally. Some even bought houses they loved it so much. Erica was proud when they filled up St Just with people who were well-paid, renting houses and spending money in shops. They came from all over the world. On the last series they had Italians, Portuguese, Spanish and a South African. A guy called Bariego, who had never been out of Mexico, arrived after a two-day journey. He absolutely loved it, got a local girlfriend and stayed for four years before the work ran out and he was forced to go to Manchester. Bariego was the happiest, friendliest guy in the world. With a big crew like that there was always a group of people going down to the beach after work. Some took up surfing. Bariego threw himself into merengue and salsa clubs. It was a warm, gossipy, welcoming small town. Everyone was so interested in them, where they'd come from. They always went to the Star for lunch. In their series they had a bar, a focal point for the town, and named it after the Star. The landlord was called Taxi Crab and had lots of jobs just like Johnny McFadden, who was a fireman, farmer and landlord. They all dressed up for the premiere in the Star.

On Monday nights in the Star there was Celtic folk music. Tourists liked it, it was packed. In winter there were more people playing than drinkers. They took their Disney executive there. She loved it. The head of the channel came. They put them in front of Johnny McFadden and he'd say something inappropriate. Once an executive from Burbank came to audit them for a week. He wanted to stay in a hotel with a gym. They booked him into one in St Ives, which he said was a cross between *The Shining* and *The Rocky Horror Picture Show*.

'Can we get you something for lunch?' they'd ask. He'd run through some requests. They'd look out the window. This was before they had a proper deli in St Just.

'You can have a pasty, fish and chips, or a Co-op sandwich.'

'Can I have fish and chips without batter or chips, just with salad?'

Erica became involved in the Lafrowda Festival in St Just. Near her in Botallack there lived a youth worker called Mary Ann. She had fallen in love with St Just when she cycled through it in 1983, with a Cornish girlfriend. She moved there the next year. The boys she looked after in the community were very close to one another, out in the middle of nowhere. They smoked dope together, one or two even took ketamine or heroin. There was very little work and everyone with aspirations, like Rikki, left. Mary Ann wanted to do something positive for these young men. She noticed the further west she went the more radical the young people's views. Asylum seekers should be shot or drowned, they told her. St Just was run by established families. The four McFadden brothers are the mayor, councillor, butcher and landlord of the Star Inn. Johnny the landlord polices any fights; the real police are nine miles away.

Mary Ann liked the fact St Just is such an anarchic, edgy place, but it had an unpleasant side to it. Cornish people are friendly on the outside, jealous when outsiders do well. On the cliffs she found a hand-painted sign: 'ENGLISH OUT'. It keeps being retouched. She started to plan a vibrant procession to involve the young people, reviving old customs. Her husband, a sculptor, made animated processional puppets of black dragons up to five metres high. They worked with the young people creating enormous puppets of mermen, Chinese dragons, leopards, cows and pirates. She called it the Lafrowda Festival. The locals hated the idea of her 'hippie festival'. Even the vicar tried to stop it.

At the heart of St Just's history is the ancient site of an amphitheatre, the Plen an Gwarry. The scholar monks of Glasney College, Cornwall's centre of ecclesiastical power in Penryn, wrote passion plays in Cornish, known as '*gwaryes*', which were performed in the round in earthen amphitheatres or '*plen-an-gwarrys*'. The church aimed to convert the peasantry to Christianity through rowdy, bawdy mystery plays, performed in Middle Cornish, about the origin of the world, the passion of Christ and resurrection. The St Just *plen* is one of only two surviving outdoor amphitheatres, out of fifty raised on church lands all over Cornwall in the Middle Ages. The plays aimed to resist the westward drift of English, to keep the Cornish language alive. In the mid-sixteenth century the west Cornish were very different from the English.

Mary Ann set out to re-enact the plays. It was a huge undertaking. Scripts were adapted from ancient texts written by monks. She involved 200 local people as actors, choir and band members, or as stage-builders to construct the eight stages. After the dress rehearsal, a cluster of locals drunkenly watched Mary Ann clearing up. One got up, walked over and started pissing on the stage.

'What are you DOING?' Mary Ann shouted.

'Pissing.'

'Why?'

'Because this belongs to me.'

She shouted, he became violent and eventually the police took him off to the cells in Camborne. The next day he hitched back to St Just. By chance, Mary Ann's friends, driving from Redruth to see the first performance, picked him up. The route was lined with thousands of people who thronged at the bar of the Star Inn. All the local cafés, pubs and B&Bs made a fortune. Once the McFaddens started selling barbecued burgers by the boxful they were happy. The roads were closed, the town was festooned in decorations and they all had a party. Over four years it brought in 30,000 people.

In 2014, Lafrowda Festival had a contest to see who could make the best fifteen-second film. Spider Eye's team helped some of the film-makers so the shorts were slickly edited and used animation. The best were screened at a 'Night at the Oscars' at Cape Cornwall school to raise money for the next year's Lafrowda. Guests were photographed by the paparazzi as they arrived in full black tie and evening gowns; they were interviewed on the red carpet, their images flashed up on an enormous screen inside the school. They were served champagne and canapés by an army of girls who ushered them to their seats. The floats in the procession had lifesize film-themed puppets: Bonnie and Clyde, characters from *Avatar*. There was a longer film by some local lads from St Just, who leapt across the old beams and rafters down in the Botallack and Levant mines. A photographer won best actor for his portrayal of Gollum. Johnny McFadden made a speech when he won his award for playing himself, a disgruntled landlord. He said that when he grew up in St Just he had no idea that one day they would see so much creativity. It was a clever speech. The audience was touched.

St Just is not a wishy-washy place. It has a strong feeling. You'll either like it or you won't. Occasionally someone says something about incomers. Clive Williams, a farmer and captain of the fire brigade, is a quiet, wise person. 'Look around you, look at Lafrowda, St Justers wouldn't have organised that. We would have thought about it but we would never have organised it. It took incomers to actually organise it.'

7

It's Christmas in Newlyn. The string of coloured lights runs all the way up to the top of Fore Street to the Red Lion, high up on the ridge, where the road snakes towards Mousehole. From the Red Lion's 300-year-old granite-fronted doorway there are sweeping views of the boats in the quay. The bulbs' reflection becomes a picket fence of light sticks on the harbour's surface; a flashing neon Santa and reindeer leap over the opening of the quay, a green tree perches on top of the harbour's new ice plant. One year the Stevensons lit up the outriggers and mast on a beamer. Transport containers are lined up, waiting for their trucks. A cab rolls into the quay with four men in overalls perched on the back, ready to slot it into a container. Because of the market the cold, brackish air reeks of fish, diesel and seawater. The last fish auction before Christmas is 19 December.

Tucked in the back of the Red Lion's split-level bar with a wild furze-thick nest of white hair, Larry sits on his stool. It's a long time since he pounded ice off the rails of a Newfoundland trawler. His body is wiry and gaunt. He has a piratical earring, a pointed goatee and radial laughter-creases around the eyes. His tiny black pupils glisten like whelks. He listens to Jackie the landlady, his head bowed. Her muttered barbs, tartly mocking the drunken fishermen she has been putting up with for twenty-eight years, make her a dead-ringer for Maggie Smith. The longest-serving landlady in Newlyn, she fiercely guards her recipe for crab soup. She has a soft spot for Larry: maybe because his wife Sarah died eight years ago, maybe because he hasn't fished for months. He is under Jackie's wing, renting a room above the bar; she tries to get him to cook his own cod in the tiny galley kitchen. If Jackie and her long-suffering husband Tom aren't there, he'll pull the odd pint. Long-furred Harry prowls down the top of the bar, brushing past Larry's pint, scouring for shreds of batter from a plate of half-picked fish and chips.

'Do you like Uncle Larry today?' Jackie coos, digging her nails into his black fur. 'Larry and Harry. He likes it because it's warm. I have to brush him. He's just a common old Gwavas Estate cat but he's really smart.'

'He's a cat and a half,' snorts Larry.

'Is he a mouser?' someone asks.

'Is he a mouser? Are you fucking joking me? If he saw a mouse he'd run! You know what they feed him on? Fucking crab. Lobsters. Prawns. Best-fed cat in the fucking country.'

Larry first met his wife on Penzance quay. He glanced up from mending his gear to see a beautiful girl with long dark hair and dark eyes, holding out a spliff to him. In the other hand she held a gorgeous grey lurcher.

'Fucking ideal,' he'd said, edging up the seaweed-coated ladder.

Her name was Sarah Hillyard. She had been at Falmouth Art College for a couple of years. She was Jeremy Le Grice's niece. Her mother was Jeremy's twin sister. She was way out of Larry's league; Larry was working-class Irish, from a small fishing village called Howth, outside Dublin; he came down to Newlyn with the Irish boats.

Her dad was Irish and she was very close to him. A keen cliff climber and walker, he took his dogs out one night somewhere in St Just, and disappeared off the cliffs. His body was washed up on the beach.

Larry and Sarah fell in love. He was a Catholic and she was a Protestant. In those days riots and petrol bombs were still raining down in Belfast, kids were being blown up in pubs and civilians gunned down by British paras in Derry. So to be dating a Prod was a real no-no.

'Fuck it, they can only shoot me,' Larry had said.

They wanted to buy a house out on the cliffs in St Just, move in together and start a life. They would be near to Sarah's family. Larry didn't want to wait for years to buy it so he needed to earn some money quickly. His friend Nicky told him he could earn big money fishing out of Boston.

'If you come over to America with me, you'll earn £1,000 a night,' Nicky had said.

'Well, that'll do me.'

It was minus twenty in Boston. Larry had never known cold like it. If they ripped the nets up Larry was the man to sew them back together again. He had to take his gloves off to work the needle, and could only do fifteen minutes at a time before his fingers felt flayed and he couldn't pick anything up. Another man unstuck his swollen knuckles with a hairdryer, kept the wind chill off. Larry's long wet hair froze.

One time, they passed Block Island in the North Atlantic and a fierce snow blizzard blew up out of nowhere, a shattering wind that froze them to the bone. The sheer weight of their catch made the boat list like a drunken man: she was dangerously unstable. Larry had to get rid of the fish, lighten her up. Snow-blinded, Larry staggered out on deck. He couldn't see his hand in front of his face. He stood aft on the net and hauled it up. The fish slid out as the net rose up. Then a sudden gust of wind made her heave to one side. The rail dipped underwater and kept on going down until cold, grey sea swamped the deck. It closed over his head. It squeezed the life out of him, like an iron fist. He had to swim off the side deck. He saw another head bobbing nearby: the engineer had been on deck with him. The skipper hurled two survival suits out to them. Larry struggled to get his feet through the holes, gasping like a racehorse. They got the life raft over the side and wrestled themselves into that. It was two hours before they got picked up by one of the coast-guard boats.

Larry felt the cold for days after that.

He stayed until he could bear the cold no longer. Sarah had waited for him for two long years. When he returned he married her and bought a house out in St Just. They had a son, Chechal, who is now at university.

Jude Le Grice, his wife's cousin, still pops in to visit Larry in the Red Lion. Jackie and the fishermen all know Jude. A group of Welsh singers were there once and Jude fell in with them. When he sang, the Welshmen nodded in wonder. What a golden voice!

Along the bar the afternoon shift sucks in the calm before the boats will spew out their wild-eyed, thirsty crews in a few hours. Nick, the Stevensons' second onshore engineer, stands behind Larry's shoulder. Nick's father was in the Merchant Navy; until Nick was 7 he travelled the world with him. Beside him sits

Terry, with trim white moustache and clean check shirt, sleeves rolled back to show bronze bracelets. He is ready to retire but is still single. How did that happen? His cheeky laugh reveals four front teeth intact, then two gaps where the canines were. Twenty-five years a skipper in Newlyn and fifteen before that in Grimsby. Came down from Cleethorpes with all the Grimmies.

They reminisce about the wild days in the Swordy. Fights. Shagging. Parties. Very hard men with more money than they knew what to do with.

'First day in Newlyn,' Larry sighs. 'Walked into the Swordfish and whack, punched straight in the mouth. My mate was going out with this broad, keeping her warm, then her boyfriend came back from sea.'

'I'm a seafarer's son,' Nick reminds him with his sleepy Wirral burr. He sounds a bit like Daniel Craig and he'd love to have his money. He never touches spirits except for one vodka and lemonade on his birthday and New Year. He went off the rails eight years ago. Now he gets to see everyone else crumbling apart. 'Only takes one spark to ignite the fire.

'I've seen the Swordfish in chaos. Bodies beating each other up. Frenchmen, Grimmies, smashed up windows. They were young, fit men in their day. You had to walk up and down that quay looking for work. There was that many bodies looking for work. Eight o'clock in the morning, you've got men looking for work.'

'Now you can't find any fucker,' sighs Terry.

'You can't find anyone to take the boat.'

'High oil prices and everything went to fuck.'

'Gone down now. 23p. 11p is the duty. 17 is the office price.'

The price of diesel is a staple topic in the quayside pubs.

The danger is that at low tide the boats sink below the quay: the only way down is on a steep ladder, wrapped in wet seaweeds and oil, slippery as a rotten peach. Not recommended after a skinful in the Swordy.

'Too many have been found in the harbour next day.'

'I've found two in there.'

'Only couple of year ago I found a 17-year-old lad. First trip at sea. Got a bit drunk. Went back to the boat.'

The 17-year-old, from Alderney, had seven pints in the Swordfish, then walked shakily to the quay with his girlfriend, who watched him put his foot on the first rung of the ladder, then left. He must have slipped on the weeds and oils, slid down the slope of the harbour wall and under the boat, where his body was found by police divers next day.

'Too many gone in the harbour. Better to say – don't come back to the boat, book into Swordy or the Lugger. Too many gone down the ladder.'

'If you're drunk and you fall in there's nobody around,' Larry said. 'Bang your head. You're fucked. Years ago on the Irish boats, we lived aboard the bloody things.

'Had to because nobody would take you in for the night, would they?'

Chuckles ripple down the bar.

'Still the same.'

Laughter erupts. They let themselves have a good roar. This is better.

One perennial favourite is discussing who is the hardest fisherman in Newlyn.

'Fish is hard, but Nigel McCrindle is harder.'

'Elsy is the hardest of all.'

This prompts another great Newlyn yarn about the time three men from the Penzance mafia came down to see Fish in the Star Inn on the quay. Fish's real name is Robert McCreath; he came down on one of the Scottish pursers, the big mackerel boats.

'I was in town,' says Terry. 'I heard them say they were coming down. I rang Debbie in the Star and said, tell Fish watch out there's three of them coming. He just carried on drinking.'

Fish was in the pub with his pal Spooky, who worked with him on the *Trevessa*, a sixty-five footer. Fish told him he didn't want him getting involved so he sent him home.

'The first one, the black curly-haired one, walked in, fucking nutted Fish. And Fish looked up and said "Is that the best you've got?" Nutted him back.'

'He laid out the second one. Then laid him out again.'

'He slugged the third one.'

'He put one through the top of the door and then the other back through the bottom of the door.'

'Then he went to town the next day to see them. He said right I don't want no repercussions. We sort this out now. Me and fucking you. And they said sorry we got the wrong man. We sent the boys over and got the wrong man.'

The cluster at the bar explodes with laughter. It's a great story and they enjoy the telling.

'He's a big lump of a man. Well, he was.'

'He still is. Got a head like a horse.'

'When he was drinking and a row started he would see red. I've seen him in Holland. I said to the barman, don't give him any more of that whisky. You can look after him if he starts on the muck. Next thing he has another bottle. There were all these Scotch men there. There were ructions! Fish was trawling them around the place like *rag dolls*. His own fucking countrymen! Fuck me. Barman says can you not do that. I said – You fucking gave him the Jack Daniels!'

Another wave of laughter burbles along the bar. This is good craic. It's still early afternoon.

'The next morning we had to pay for a load of drink, a telly and fucking drinks cabinet. Fucking unbelievable.'

One person who was never going to win the crown of Newlyn's hardest man was Grimmy Brian. His catchphrase was: 'I've had 364 fights and I've lost every single one of them.' He wasn't even from Grimsby but Whitehaven, 200 miles away. In the Star there is a creepy, shiny bust of a moustachioed man which looks exactly like him. Brian nicked it from a charity shop. In the Red Lion they all have stories about him.

Once he rang a taxi and said, 'Ten minutes.' For some reason the taxi turned up early, which is unheard of round Newlyn. Brian hadn't finished his drink.

'I told you ten minutes,' he said holding up both hands.

'I can only count seven fingers', the cab driver replied.

Brian'd lost his fingers at sea and had to retire. He was sitting with four men round a table in the Star: Dealy, Vince, Spooky and another.

'Do you realise that not one of you cunts has ten fingers,' someone said.

'A lovely man,' Larry said. 'A beauty he was.'

'He asked to borrow twenty quid,' Terry said. 'He was dead the next day, fucking bastard.'

They chuckle. Then stare ahead with cloudy eyes.

'He had a fetish for women's underwear,' Larry said. 'He loved wearing it. He didn't mind anyone knowing about it.'

'Even the vicar said that at the funeral. The notes were handed to him. He didn't read through them. Then he read out to the whole congregation that Brian preferred to go to sea with women's underwear on.'

'I've seen him one night with the barmaid, had her knickers on in ten minutes. She went out, gave him her knickers, he fucking put them on.'

'Barmaids used to go round charity shops to get knickers, he thought they were theirs but they weren't.'

Amidst the nautical photos on the wall of the Star, there are several of Brian wearing full drag.

Next to Wirral Nick sits his boss, Paul Strowger, chief engineer for the Stevensons for forty-two years. He started working for them when he left school in 1972, first on the fish market and then in engineering. He is broad-shouldered, with receding hair and a deep frown line dug between thick black eyebrows. He remembers when the Penzance chapter of Scorpio bikers came down. Squeeze, a sinewy fellow covered in tattoos who does the cellar in the Star at ten every morning, used to be one of the main men in the biker clubs in those days.

'They thought they were hard and came down to Newlyn, but the Newlyn crowd outfought, outdrank and outdrugged them.'

In the mackerel boom the fishermen all had motorbikes. Fish and Spooky had them. There was a bookie's right next door to the Swordy. When the fishermen had lost their money in it, they would try to trick it off the others with three-card brag. Nick claims that they once strung a guy up in the Swordfish, doused him in lighter fuel and stuck a match out the top of his hair and swung him to see if they could strike the match on the floor. No one believes this for a minute. It's a yarn trotted out for the tourists.

Larry's done more drugs than the man in the moon and it hasn't fucked his head up. Not that he knows of anyway. If it did the whole town would be in trouble. There are loads of them in Newlyn all off their heads, although there's not a major smack problem like he has seen in places like Dublin.

'They don't do it at sea. That's your rehab out there,' Larry says. 'That's your fucking rehab. You go there to get away from it all. If you work your bollocks off you don't think about it. Party party, get wankered for three days, brandy, get stoned. Going to sea is rehab.'

'We call it the Bet Ford Clinic,' Jackie smiles. 'If they do two trips then take the third one off, they are home for ten days then. That's when merry hell starts happening.'

'Coke heads, gear heads, dope heads,' Larry goes on. 'Once they go out there that's it. If I'm off for a week and I go out there I'll be sick as a fucking dog for the first twenty-four hours then after that I'm alright.'

From nowhere the portly local woman vicar looms behind them, grinning like a baby with dimpled cheeks; she offers them blood-dark plums from an open tupperware box. Surprised to find some healthy solids, they post them in their mouths. Larry nearly chokes on one: the plums are so laced with brandy that it's like eating a rag soaked in paraffin. She must have left them in the bottle for years. Whether it's the brandy or the Kraken black-spiced rum that is now doing the rounds, things steadily start to deteriorate. At the mention of one skipper's name an argument suddenly flares up out of nowhere. This skipper shagged another fisherman's wife.

'Don't trust your best mate!' Terry shouts, swiping the air with his arm.

'Don't trust your best mate,' Strowger echoes back, tossing his head right back like a horsefly had bitten his ear. They swat the other's words away. Blinking slowly, Strowger lets his mouth lie open for some time as he prepares his next outburst. Larry tries to butt in. He and Strowger were in a hurricane together, for fifty hours, heading to Vigo, facing the Atlantic on the north-west tip in Spain. It was so rough that you couldn't even have an inch in the kettle or it'd spill out.

The shouting ramps up. Swearing. It's kicking off early for three o'clock.

'Peace, love and butterflies!' Jackie yells. She has to do something. So she cranks up Slade's 'Merry Christmas Everybody' to drown them out, clapping her hands above her head as if she's at the concert. A few more months of dealing with drunk, scrapping fishermen then she'll be on holiday in Thailand. There's a whole group going: some of the fisherman have come back with Thai brides who've then left them.

She's a bit worried about Larry. He's been on edge since he opened an official-looking letter earlier, addressed to him, care of the Red Lion. He read it several times nervously jiggling his

211

leg, rubbing his upper lip. It was from the Marine Management Organisation who set the quotas. They were enquiring about a couple of missing boxes of fish from a trip months ago. A black cloud had gone over Larry's face.

'From tally-up, the fish's gone missing somewhere. I don't fucking know. Said he wants to have a word with me about it. I don't know,' he sighs. 'Made an appointment to go and see them. See what they say.'

Larry shuffles out for a smoke through the swing door. Jackie feels maternal about him. She put a cat, Milo, up in his flat for company, but Larry came in pissed, trod in the litter tray, trod in the cat food and then flung the cat out soon afterwards. Now this MMO note is making him so stressed out she's worried he's going to have an asthma attack.

Larry comes back from his smoke and sits on his stool unable to speak. The others chat for a while before they even notice he's back. He is bobbing from side to side on the stool like a dan flag. Not in time to the jukebox. His eyes stare ahead, pinned like he's lost in a sea fog. The tall, thin stool wobbles.

'Larry's at sea!' shrieks Jackie, clapping her hands in delight.

The following day the boats are in. Down nearer the quay the Star fills up with men in their oilers. They've been at sea for days and are now gasping for a pint. Two of the old-school Newlyn crowd, Perry and Vince, sit in caps and their oilers, hunched over photos of their kids on their phones. Perry is a giant, six foot four. His shoulders are as broad as outriggers. His hands huge as shanks of lamb. He shouts out the side of his mouth like Popeye, thrusting his face forward. Every shouted word

sounds like a threat. People steer clear of him in the afternoon.
To the fishermen he's a pain in the arse, but a loveable clown.
To the unacquainted he's an aggressive, dangerous drunk. If
anything needs to be sorted out, they make sure they talk to him
in the morning.

Perry is chief engineer on the *St George's*. The boat has grossed
over a million this year. He's been skipper for a lot of the boats.
He came down when he was 17. He had really long hair like
a Celtic chieftain, rode a motorbike right through the Swordfish.
He called himself 'Warrior'. At 24 he went to the Democratic
Republic of Congo as bosun on a survey ship with the merchant
navy. Mike Collier was the skipper. The Congo was, at the
time, a haven for mercenaries, diamond smugglers and money
launderers. It was run by a despot leader in a cowboy hat, so
corrupt he had bought himself a Swiss bank to hide his billions.
Mike was used to Africa. His chief engineer, a keen gun man,
even showed the local soldiers how to strip down an AK47 and
put it together again. One night Mike threw a lavish party on
board for some French colonials he hoped would charter their
boat. Perry took it upon himself to let off a fire extinguisher
all over the guests. Mike sacked him. Perry ran at him and
drop-kicked him in the back, breaking two of his ribs. Then
he took off on a motorbike with an Irish friend through the
Congo's rainforests, encountering mercenaries on the way. Perry
claims the colonials were abusing kids in their luxury hacienda.
Mike firmly denies this.

In Perry's mind he was always the avenging 'Warrior' fighting
for a noble cause. He hated bullies. He once climbed on the
Twilight, rail pinned a fisherman and stove his head in, because

he had beaten up his girlfriend. His father was a marine who used to beat his mother. 'If you beat up women I will shoot you,' Perry said. He knocked out a jealous skipper who came round his house looking for his soon-to-be-ex-wife, when Perry's son was only two weeks old. He hauled Baden Madron outside the Star and knocked his teeth out, because Baden came out the toilet, pissed out of his head, and bowled Perry's pal Helen over. Helen made jewellery and was arty, bohemian. Her closest friend kept a horse in her living room.

'There's no excuse for pushing a woman over,' Perry said.

Perry's father beat Perry as a kid too. When Perry grew strong enough to defend himself he walloped a guy with a frying pan and was put in the detention centre with his hair shaved off.

Perry's warlike temper came back to haunt him when Mike Collier turned up in Newlyn as the Marine and Coastguard Agency guy, checking to see if the boats were seaworthy on behalf of the Department of Trade and Industry. His checks were not popular. For six months no one spoke to him.

'What do you want?' they'd say.

'A cup of tea.'

Once on board he'd check the kettle, then he'd do a Lieutenant Columbo double take just as he was leaving, nip back and check the fire extinguisher, nautical almanac and everything else. They tried to trick him by swapping the newest items around between the boats. If someone had a good set of flares or brand new charts they would switch them fast onto the boat just before Mike came on board. So Mike started numbering them all with a permanent marker. One day he went on board the *Filadelfia* and met up with the engineer.

'You broke my ribs in the Congo,' Mike said. It was Perry. It was pretty awkward. Mike Collier didn't hold any grudges though. He later became harbour master in Mousehole.

It was small like a village. Perry's ex-wife married Baden Madron's brother, Shaun. Then Baden married Stacey's ex-wife. They all went to the wedding at Lanyon Quoit, a megalithic tomb from 4,000 BC, two miles south of Morvah. It has two standing stones and a larger flat capstone balanced on top like a table. Guests wore Cornish tartan. The Scorpio bikers ran the bar.

Vince is Perry's unlikely partner in crime. They go back. Vince is 45, tall, handsome with dark grey hair and a clean white smile. But there is a bristly restlessness to him. He's as unquiet as the sea. His eyes are weary and spent. He's thirteen years younger than Perry, who is nearly 58. Maybe the old warrior felt protective of him, because Vince's uncle, Roger Nowell, had saved Perry's life. Roger, who was Newlyn's best-known and most-loved skipper, sailed with Perry. Every four hours they shot the nets. Perry was on deck. As the nets slipped over the side he caught his foot in the rope tied to the bottom of the net and was swept overboard. Snared by his ankle, he was dragged along eighteen feet under the surface at a speed of five knots. Terrifying. As the boat surged ahead at five knots, he was towed by his foot, the rest of him streaming out behind like a waterskier. The force of the current meant his hands were trailing behind his head like river reeds. Above him he could see a mass of bubbles from the net being dragged through the water. Any other man would have drowned. But Perry was very strong. He twisted and writhed against the surge of the current and managed to pull himself up, then clawed his hand along his leg, gripping his clothes in fistfuls until he was bent double and could grab his foot. Luckily he was wearing Roger's sea boots, which were one size too big for him. As soon as he released the tension of the rope he could wriggle his feet out of them. With lungs bursting he burrowed blindly up, legs kicking in a frenzy, until he broke the surface and gasped. Another moment underwater and he would have inhaled seawater. As he heaved in the air he scoured around for the trawler. A new terrible thought came to him: they had not even seen him go over. The boat would keep going, further and further away. Then something extraordinary happened.

From the water Perry saw this huge steel hull bearing down on him. It slowed, came alongside and they grabbed his arms. They held him against the side of the boat, keeping calm. Not making a big deal out of it. After he got his breath back they hauled him on board. They took him aft and put him under a warm shower. In the galley with a mug of tea, he told them what happened.

Roger had been in his bunk when the engineer came in shouting, 'We've lost Perry.' He ran to the wheelhouse and saw a body disappearing in the water behind them. A deep swell was coming and Roger knew he had to keep his eye on him or he would never be seen again. He stretched out his arm and pointed. The mate turned the boat around, steering towards the line of his arm. Perry had survived some terrible scrapes. He'd bust every bone in his hand. Now his warrior days were over. A few months ago Perry had got into a scrap in Penzance. They beat him so badly he was off work for seven weeks.

'The trouble with Perry,' Vince says, 'is that he still thinks he's 25.'

*

Winter is a dangerous time. The storms always come.

On Valentine's Day in 2014 a terrible eighty-mile-an-hour gale struck Newlyn, one of the worst in years. A surge sent a wall of water vaulting over Newlyn Bridge, swamping all the shops. Twenty homes were underwater. The wind sliced through cables, leaving thousands without power. The fishermen sat the winter storms out for ten weeks at home. Only a few of the smaller boats went out. They had to. The bank was after them for repayments; their tax had been due at the end of January. They were gill netters heading for deep-water wartime wrecks, a hundred miles out, where pollock congregate to spawn. Patch Harvey, the coxswain of the Penlee lifeboat, knew them all. He followed their blip on the AIS tracking system for sixty miles, then they were gone. The cost of diesel meant they'd have to fish twice as many wrecks per trip and risk being a hundred miles off when the weather turned.

Patch had a share in a gill netter for twenty years. He knew how you shoot the gill nets over the top of wrecks in winter to try to catch the pollock. It's like throwing a parachute over a church. If it snags it will rip the thin monofilaments to shreds, but you can get three grand from one wreck site. They can only work the neap tides: a fourteen-day window to make money. The pull of the moon is too strong for gill nets in the spring tides.

Patch was trapped by a bad winter storm once, a force ten with sixty-mile-an-hour winds. The sea was white. Wind was howling. A thirty-feet swell was breaking over the bow. He struggled to keep the boat from going side-on to breaking waves that would roll her over and flood down below so fast she would never roll back.

There was the time Michael Fish said there was no storm coming. Some of the Newlyn boats were off the Scillies. From flat calm it blew up into a hurricane. They had never seen weather like it. Sixty-feet waves. The bow was lost in the spray. You could be fifty yards from another boat as it disappeared into the depths of a trough. Horrible seas. Fishermen who'd only worked the North Sea couldn't believe it. The Atlantic and the Gulf Stream, together with the continental shelf off Ireland, created devastating surges. The strong tide runs one way, the weather comes against it. The swells are longer. At times like that men can sink to their knees and pray. You have to be spiritual in fishing to get through that weather. They trust their lives to men like Peter Ellsworthy the skipper of the *Cornishman*. Clever man, knows the tides, the seas. Ninety per cent of the skippers in Newlyn are top-notch.

Allie, skipper of the *Stylissa*, one of the top boats in the port, fishes in storms: he goes out in bad weather hoping for ten grand a week. His highest gross is fifty-six grand for one trip.

'I've got five kids. I've not had a holiday with my family for three years,' Allie tells Perry. 'I haven't had a honeymoon with my wife yet.'

'I've got the port record,' Perry says. 'I burn 3,000 litres of fuel a day.'

'We burn 800 litres a day.'

'You just dodge, we tow.'

'We steam around.'

'Steam around a fucking set of meat.'

'Put it another way,' Allie says. 'Every time they are banging money out the fucking mission we never missed a fucking trip this winter. When the weather broke and everyone fucking got out and the prices went to shit, we were stopped from landing. I sold a whole fucking trip, 330 fucking boxes for 80p a kilo. Those cunts.'

'Egg and chips,' Perry says.

'We've had plenty of egg and chips. And more yet. We've worked weather we should never have done. Pulled in some . . . '

'If you're skint I'll give you fifty grand.'

'I'll have a fag off you.'

'We work storms.'

'What's diesel at the minute?'

'Fifty-one.'

'My gear is twice the size of his. And my dick.'

The Atlantic winter is a time when Newlyn share fishermen can make a killing. Hardly any boats risk going out which means fish prices are sky high. But they have a fight to stay out in the worst weather that can become a fight to stay alive. They go through the gaps and don't know if they will come back. That gives an edge to the pints they sink in the Swordfish and the Star.

'I'm going up the Red Lion to eat,' Vince says. It's near where he lives. He can have a smoke with Larry. Perry hesitates. Jackie is not his biggest fan. He is more welcome in the Star. The barmaid has taken him home when he's been drunk. She even knows his pin number. Just by the door in the Star is a board marked: Perry's Knocking Board. He can play the spoons there when he's drunk; they put it up to stop him knocking all the plaster off the wall.

They both squeeze into Vince's second-hand two-seater MG bought in September with his summer wages. Vince's second wife, Juliana, pointed out that a two-seater was no good for carrying their two very young kids. They have been together seven and a half years. They split up three weeks ago. She's moved into Penzance with her mum. He has two older kids from his first wife. He keeps a five-bedroom house, but it's empty. The heating's off.

He's been working abroad, fishing in Egypt and on the standby ships in the North Sea. So they've grown apart. She was a crab picker when he met her. She looked gorgeous in a red bikini. They went out for a swim. She grew up in the Ukraine, but moved to Lithuania. Vince had to pay off the Lithuanian police. When he worked out in Egypt he realised that everything is baksheesh. Everyone has their price. Most of his crew got their tickets from baksheesh.

Vince is a complicated character; he's one of the few skippers with a class-one ticket, which shows he is highly competent. He is one of the most intelligent fishermen in Newlyn. He has a brain and a half on him. He is proud of his Cornish roots. Vince reads Crosbie Garstin's *The Owl's House*, a swashbuckling epic of smugglers, press gangs and pirates. He reads it for the Cornish dialect and remembers things his gran used to say, that he doesn't hear now that the Cornish language is not widely spoken. There's been growing interest since the 1980s and about five hundred people speak fluent Cornish.

Vince is captain of the *Elvery*. He spends more time with his crew than his real family: they have become his family now. They clean their boat up, do housework. That's where they live. They don't just fucking fish.

'We are real fishermen,' he tells Perry. 'Hard core, not plastic cunts, not Mickey Mouse motherfuckers who . . .' he trails off.

Up on watch at three in the morning the sea is desolate, slate-grey in the moonlight. He plans all the good things he is going to do when he comes ashore. Thinks about everything he's done that made him who he is today. He thinks about his kids with his first wife: how they grow up and move on but he is always concerned about them. His daughter is 17; his son is bigger than him now. He thinks about Juliana and his younger kids with her. Then comes up with plans of how he will turn things around and make it alright again. But then he comes ashore and goes into the Star and it all goes to shit.

'This is what happens when your missus buggers off,' Vince says as they head into the Red Lion looking for Larry's favourite corner of the split-level bar. 'You have to eat out the whole time.'

But something is wrong. The day before Larry had fallen off his stool, smacked his head hard on the floor and been rushed to hospital. Now he sits there looking frail.

'I had blood all over my body,' Larry croaks. 'I thought I was dead in heaven.'

'We blame it all on the vicar,' Jackie says. 'She gave us those brandied plums.'

'Every one of those plums laced with fucking brandy. I had a glass with rum in it. That fucked me right up. I was hammered. I fell off the stool that way. Me legs collapsed. I was in an ambulance.'

'Bleeding like a bastard,' Vince says.

'Fucking glued me head back together.'

'That's what they do today, they superglue it.'

'Wires and fucking heart things everywhere. Took out the blood.'

'You woke up after the blood loss and the dizziness.'

'Last thing I remember was sitting here drinking.'

Larry shakes his head, pads carefully behind the bar, serves a round. Then Jackie reappears.

'Larry!' Perry shouts at his back. 'Did you get pissed up and fall off the fucking stool you wanker?'

'First time in fifty years.'

'Yeh, you fucking lightweight.'

Vince and Perry had such long hair they were always being mistaken for each other. One night Vince was invited to loads of raves across the whole of the West Country by a total stranger who said: 'Great to see you again Perry.' Another night out raving with Brixham fishermen, they popped twenty pills, stripped off and Perry threw a bottle through a shop window in Camborne. But Vince was the one who spent the night in the cells.

'You let someone else spend the night in the nick,' Jackie sneers at Perry. 'You let your best mate take the rap for you. You've gone right down in my estimation.'

'Fuck off.'

'Right down in the gutter.'

'I PHONED UP . . . ' Perry shouts.

'DON'T SHOUT!'

'He didn't let me take the rap,' Vince says. 'He phoned up next day. I was released.'

Jackie is already moaning to another punter about Perry, who sits a few yards away, brandishing his empty glass at her.

'I said to Larry he always shouts, shouts, shouts. He has to be the centre of attention. He has to be the naughty boy. If he doesn't behave . . . I'm going to shove this where the sun don't shine. I'll ban him again. It'll be ideal.'

Perry senses the time has come for him to play the spoons.

'Jackie, could you do me a favour?'

He calls her three times before she turns. 'Could you get me some spoons?'

'You only live round the corner, go and get your own. After you've played them, people catch their lips on them.'

Vince announces he's going out for a spliff. Larry follows him.

Perry is off, lurching through the bar. He bellows at a young crowd round the pool table to put the Bruno Mars track on. He doesn't know the name. He collars several others asking if they know what the song is called.

'Perry, I honestly haven't a clue what you're talking about.'

Perry clamps his arm around one of the young men holding a pool cue. They all look up grinning nervously.

He tells them how one night, after drinking Jameson's, he shinned up a telegraph pole, jumped off it and stumbled onto the bus. His pal cut his shoe off. The foot had swollen up. He'd busted everything in his leg and couldn't walk for thirteen months. He was looked after by the man who ran the chip shop in Mousehole. He had the best time of his life, totally skint. Ended up marrying his daughter.

Vince returns from his smoke. At the bar they are passing round a book of photos of Newlyn trawlermen, *Salt of the Earth*. It is raising funds for the mission. Vince flicks through the pages.

'I don't know half the people in here,' Vince says. Then he turns the page and staring up at him are three faces he knows well: the Nowell brothers. The Nowell family have fished the waters around Newlyn for 200 years. They worked on Stevenson boats until 1988 then bought their own.

Vince's head lolls forward, sinking inches away from the photograph. He flips a few pages of the book. There is a photo of Mike Nowell's son, Jack. As a kid Mike bribed him to play rugby by paying a quid for every try he scored. One day he scored twelve against Redruth. Now Jack plays for England in the Six Nations. Both father and son have tattoos sleeved up from wrist to shoulder of their left arm – like the turquoise scales of mermen. Roger Nowell was held at gunpoint by a drug-crazed fisherman. The incident is known locally as the 'Siege of Newlyn'. Uncle Frank Nowell carried a gun on board his boat. When Perry was divorced his missus told him to get rid of his shotgun. There were a few guns around.

Vince's head is now touching the pages of the book. He's totally out of it.

'Look at Vince. Gouged out,'

He's got a case pending for beating his neighbour up. He goes to Truro Crown Court on 11 March. The grievous bodily harm charge is dropped to a single count of assault, occasioning actual bodily harm. Vince pleads guilty. He gets a 2-year suspended sentence and has to pay £1,500 compensation.

'The reverend fucked me up,' Larry grunts and disappears off upstairs, blaming a headache. Vince is passed out, head resting on the bar. Nick, deputy engineer, who doesn't drink, decides it is time for him to leave. He makes it as far as the jukebox before Perry looms in front of him.

'This man here is my chief engineer,' Perry shouts out. 'He can strip you down and rebuild you.'

There is laughter.

'See you again cap. You take care boy,' Nick says.

'I've been out in hurricanes. Over the side. I've lost so many of my friends at sea. Saved a few too,' says Perry.

'You have. Mickey Merriman, you pulled him off the rail.'

A freak wave had come in, a wall of dark water that rushed up behind them. It picked Mickey Merriman up, took him away. Perry climbed up the mast so as not to lose sight of him. They did a figure of eight, standard safety procedure. Mickey was in the water. Perry used his brute strength to pull him back onboard.

'You saved him. You pulled him back. On the *Adelphi*.'

'Bullshit. We was on the *Dunwaters*! We was on the *Dunwaters*!'

'Well, there you go.'

'She was lost off the North Sea. With all hands.'

'When they get the bodies back out the trawler . . . '

'My mate Robbie, he had his tongue hanging out like that. He'd been down. Young Roger Nowell picked him. I've seen a lot of men drowned at sea. Salvation.'

'I've got to go now Perry.'

Perry has clamped his hand around the back of Nick's neck, drawing him in close. Nick puts his arm on Perry's shoulder, reassuring him. Perry's eyes roam the floor, desperately looking for something. He chews his lip. He is so tired. The fatigue is the worst part. He looks up at Nick. All around them is mayhem and shouting and drunken yelling. He senses he has to end on a cheerful note.

'I've got the biggest engine in the harbour,' he says. 'Forty-five litres. Four-ton flywheel. A motherfucker of an engine.'

Nick listens carefully to him, like a patient father.

'When I lose the plot of my life I'm going to pinch her and take her to America.'

*

Patch Harvey remembers when he sank with Nutty Noah. They were in the water in full oilers, boots, tops.

'I knew the sea would take me one day,' Martin had called out.

Patch swam over to the life raft. 'Come on you daft bugger, get in with us,' he said.

They all liked Martin in Newlyn. He was a one-off, a bit mad, but one of the top crab fishermen in Cadgwith.

Patch's dad, Edward Harvey, with his long hair and cap, is a real Newlyn character, one of the last original trawlermen left. He's been at sea all his life. They wouldn't let him go out as a boy until he could splice wire and mend ropes. He was a young lad when he first went to sea with the Stevensons, and he ended up a skipper for them for twenty-five years. He's 70-odd now, still fit and strong, up at six every morning, goes down to the boats, helps himself to a fish, fillets for the rest of the crew, has a drink in the pub with them. He's a bit of a loony, but he helps repair nets and splice wires because the crew these days can't do it.

He can remember all the fishing grounds and wreck sites in his head, sail out of the bay without any gadgets or coloured plotters, shoot his trawl and turn around.

After twenty years Patch retired from fishing. He joined the lifeboat and sold his share in the *Sabre*, a thirty-feet netter, to his old partner Carl, who had also been on Nutty Noah's boat the night it sank. Patch's oldest friend Mark had retired too. A popular guy, with a smile on his face, Mark could never walk past a struggling fisherman, so he helped repair vessels on the harbour front. One day, in March 2004, Mark crewed for Carl on the *Sabre* when he was a man short. At midnight a call came to the lifeboat station. The *Sabre* had sunk with three hands. Patch raked the seas in the darkness. He was godfather to Mark's kids. Mark's wife was pregnant. After twelve hours Patch found them, drifting, off the Lizard. Mark had drowned. Carl, who was fit and strong, had hauled the third crew member on the life raft, then heard Mark's cries for help grow fainter in the darkness. It was pitch black, freezing cold. They were battered by fierce winds and waves, not knowing if their wives had raised the alarm. When they set sail it had been a beautiful, calm day. Carl never went to sea again.

Patch is proud to be coxswain of the Penlee lifeboat, the most famous lifeboat in Britain. Like his father before him, he has been a crew member on the lifeboat for twenty years. Older fishermen drop in, steal his biscuits after hand-lining bass near the Runnel Stone, by Land's End or the Mopus Ledge off the Lizard. When the ground sea cracks, they know the rocks are below. One man tells of a freak thirty-feet wave that rose up out of nowhere, and sent him upside down, broaching in a death roll.

The storms in the Atlantic generate massive waves. They build up in big depressions and the tide pushes against the wind to make the waves steeper. They hit the Cornish coast, pound Penzance's Jubilee Pool, drag the paddling New Year's Day revellers off the beach. Patch's £2 million, 2,500 horsepower lifeboat is totally sealed, so the water runs off it. His crew are all experienced, first class. One is the grandson of the Penlee crewman Nigel Brockman, who drowned in Cornwall's greatest tale of bravery, the *Solomon Browne* disaster. All are ex-fishermen who wouldn't hesitate to jump across onto another boat in the dark.

Patch likes to hear the fishermen reminisce about the old days on the Scilly Isles, when they left their gear at sea for three days while they moored up in St Mary's. Local farmers brought them rabbits and fresh food. The Scillonian Club served hot dinners through their portholes. They gave them crabs in return. They took a bucket of fresh crabs up to the Bishop and Wolf Inn, left it outside and people helped themselves. Beer flowed all night. At sea the Culdrose helicopter wound down a basket, the fishermen filled it with crabs and the pilot lowered down Royal Navy cigarettes in return.

They laugh about Ross Vickars, a local troublemaker who broke into the football clubhouse and made off with the RNLI collection box. Two months later he found himself on a sinking sixty-footer off the Scillies. He was winched to safety by the Sea King helicopter as his boat sank.

Five times a year some City moneybags in a thirty-feet yacht will set sail for a pub lunch in the Scillies, round the Lizard, hit bad weather and need help. Patch could give the skipper a rocket for endangering his crew. Most times he bites his tongue though, tows them in and they don't even thank him. No donation, nothing.

Marie Thérèse jokes that Nick Howell is no longer interested in pilchards, he is preoccupied with restoring vintage cars. At 63, he's semi-retired. He bought a 112-year-old Toledo automobile powered by a steam engine: one of only fourteen left in the world. It was a rickety old boiler on wheels, lit up with a pilot light which took half an hour to heat the steam. Marie Thérèse had a terrifying ride on it in the London to Brighton Veteran Car Run, when it burst into flames. Undeterred, Nick took it to America and drove sixty-five miles from Flagstaff to the Grand Canyon with his brother, both in period outfits. They won awards. Jay Leno, the TV star, was their gracious host.

Billy Stevenson still bristles at the mention of Nick. Rumours spread in Newlyn that Nick got an EU grant to do up and sell off the Old Pilchard Works as flats. Nick laughs this off. He still owns the bleeding thing. He's had it for thirty-five years. The EU killed his business off. New regs said all salted fish had to go in a separate chill room. All the tiny, family-run Italian grocers couldn't afford that. So they stopped buying salted pilchards. Nick's sales dived from 2,000 barrels to eighty-two. So that was the end of 500 years of selling pilchards with their head on, gut in, in wooden barrels. That was the end of the museum and the last remaining working Cornish salt pilchard factory. He turned it into flats.

What people did want, though, was Cornish sardines in a steel tin with a decorative painting on it. The market he and Nutty Noah helped create grew. It became a multimillion-pound fishery. Nick still produced tins in Brittany and continues to sell them in Waitrose and delis around the country. Out of Newlyn a twenty-first-century fleet of expensive high-tech ring-netters were built or adapted to follow in Nutty Noah's footsteps. Elizabeth Stevenson's husband Sam fished successfully for pilchards out of Newlyn in his catamaran *Lyonesse*, named after the sunken city. He supplied Tesco who sold them as whole Cornish sardines over the counter. In 2010 several new ring-netters were fitted out. They were made of fibreglass. They fished to order. Their dashboards gleamed with modern electronic navigating and fish-finding equipment. The inside of the wheel-house was lit up like a space capsule.

Nick hadn't really known whether ring-netting was going to work. Nobody had known. They all laughed at Nutty Noah and his crazy ideas: a twenty-four-footer going round in a circle and winching up the bottom of the net?

'Keep everything to yourself,' Nutty's friends had warned him. 'Everyone will copy you.'

But he had wanted to share his ideas. He even drew them pictures. The modern fleet of high-tech, fibreglass ring-netters nosing out of Newlyn now all use his techniques.

Nutty Noah stirs powdered milk in his tea. His hair has gone white and feathery like a dandelion clock. His neck is swollen. He has just heard a good mate, Gary, a Newlyn fisherman, got drunk on his birthday in the Swordfish, wobbled out and drowned in the harbour. He didn't have a home and was going back to his boat for a whisky.

Nutty is forging ahead with new ideas; full of undefeated optimism. So many ideas he can hardly keep track. His latest thing is looking for truffles. He made a special truffle spade, by cutting down a Cornish spade with an angle-grinder. A Cornish spade is wider and flatter than usual. He trained his dogs to smell truffle. Sally hid one away in the long grass, came back and the dog raced off and found it everytime. Truffles are bloody hard to find. A friend went on the computer and told him a farmer was harvesting a hundred kilos in a ten-acre field. The truffle in his fridge has gone a bit mildewed. He hasn't a clue how to cook it. Has never eaten one before.

Nutty Noah has filled every bit of space in his driftwood house with paintings. What he needs now is a separate studio. He has started building a stockade out of whole tree trunks twenty yards off, in the long grass of the field: here he can hold private views, away from the caravans. He laid a ten-feet square of concrete.

He had new knees in 2011, but years of fishing have given him a strong core. One at a time, he shouldered fully grown fir tree trunks through a gap in the hedge. He stripped the bark off, connected them together with steel bars. Three high for the walls. He's run out of money now. It cost £1,000 to put the fucking concrete down. Fuck knows how many tons of rocks were under there. Sally doesn't want him to carry on as she thinks it will fuck up her plans to put a log cabin where the caravans are. Bloody caravans. They cost £500. There's no insulation. They are cold and noisy. They're not fucking houses they're tin sheds.

He's handed out business cards for 'Cadgwith Cove Cabs' all over the Lizard. Someone has been scribbling out his mobile number. He suspects it is 'Lizard Man', a rival taxi driver who is guarding his patch. But he can't prove anything.

On Fridays, in a bright short-sleeved shirt, he walks down the steep lane to Cadgwith Cove Inn. Toots grins and waves from outside. They step over wet Labradors, squeeze in next to tourists in fleeces, tired from the coastal walk, who cradle a Betty Stoggs and wait for the singing to start. He joins the singers with his dark-toned bass. He has his arms round Matt in a Boston Red Sox baseball cap who knows all the words because he's been coming for four years. Matt has a pink chit from his wife, a mother of three, to see an old friend from Oxford who was born in Manaccan, a village on the Lizard. His friend is in a striped rugby shirt, saying how one real Celtic phrase came from Cadgwith: *quilik*, meaning bloated like a frog. They were up drinking single malt until eight in the morning with the publican in Manaccan.

There is a tap on Nutty's shoulder. It's the blonde barmaid.

'Someone is here to see you.'

An elderly lady in a blue tea dress wants to say hello. He goes over. He has time for everyone. She's Australian, with Cornish mining roots. He orders a huge round. He carries a jug of Atlantic ale and fills up the singers' glasses. All the time singing the bass. He cheers up the artisan blacksmith from Shropshire who retired from his architecture practice; he and his wife are escaping the empty nest. They all want to know Nutty. For a few hours he is a celebrity. Cadgwith is their 'home from home'; the tourists love their quaint cottage, its sea view and parking space. The singers launch into 'This is my Cornwall, and I'll tell you why/ Because I was born here and here I shall die.' But how many were born here? Even the bearded guy in the hat has only been here twenty-four years. That's how it is.

There is no mobile reception in the cove. You have to pay 50p to use the pub phone. Martin goes outside to smoke then pokes his head through the sash window and barks a request. Toots joins in if she knows the verse, and always sings like mad at the chorus. All around the pub walls are photos of Martin, Plugger and the other fishermen. Sea lanterns and rope hooks hang down; huge canvases of old men in oilers on the wall. Nutty wrote a song about his grandfather, a pilchard fisherman, who anchored a big net to the cliff for days. One night he sang a solo, dressed as a fisherman. The whole pub erupted. A local said he should go on tour with it. He's saved up some money to make a CD. And he put a link to him singing on YouTube on his taxi-cab card. He'll edit it properly when he gets a chance. In the evening he leans back on the scruffy dog sofa in his shack and looks up at his painting of seine fishing clipped to the tarpaulin roof.

There are eight fishing boats down in the cove. The last in a line that stretches back thousands of years. Mark on the *Starlight* is Nutty's age, the others are younger. 'Tonks' Tonkin and the Michel brothers are Cornishmen from the Lizard; their families go back for generations. They look out for each other.

Nutty's latest project is consuming him.

He cannot afford to go on holiday, so he went camping on Kennack Sands round the corner. In the moonlight he dug out the dunes to create a flat area for his tent. About six feet down, his Cornish spade made a zing noise. He thought, bloody hell, maybe it's a treasure chest. He looked down and saw a thick steel rod, pointed at one end. He guessed these rods were used to probe into the sand when you couldn't dig any further. Nutty decided it was a marker left behind by someone else. Nutty was well aware of the local folklore: a notorious pirate, Henry Avery, a celebrated buccaneer was being hotly pursued by the king's ship from Coverack to the Lizard. He had made for shore and buried his treasure there. His bounty amounted to twelve chests containing emeralds, topazes, rubies, sapphires and diamonds. Another chest had gold bullion. The estimated value could be £30 million. It was buried deeper than fifteen feet.

There are rumoured treasure sites in coves and beaches all round the Cornish coast. It could just be folklore. The most successful pirates were the Killigrew family in Falmouth. They owned a key piece of land which looked out over the approaches to the whole harbour. They controlled all the ships in Carrick Roads, the drowned valley, the deepest natural harbour in Western Europe. They sent their ships out all over the English Channel, the north-east Atlantic and the Irish Sea. Their home,

Arwennack House, was fortified, and the spoils of their raids stored there. Mary Killigrew enjoyed the thrill of piracy the most. She buried treasure in her garden. When a Spanish vessel anchored in Falmouth, she ordered her servants to seize the ship and its cargo.

Nutty wanted to find out more about Captain Avery's loot. He made some enquiries: a historian had been looking for treasure at Kennack Sands. Another man in the local map-and-chart department had retired early for an unknown reason. That seemed suspicious to Nutty. Nutty took a divining expert down to Kennack Sands.

He stumbled on a Daniel Defoe book called *The King of Pirates*. Defoe was supposed to have interviewed Avery. This was a book of the pirate's letters. A theory emerged. One of Avery's sloops was lost in the dark when it returned to England. Some crew made it ashore. They could have holed up on Kennack Sands. The beach was privately owned. Cornwall was pretty lawless in those days. Free trade came into Cadgwith Cove for Cornish landowners. Then Customs and Excise built a blockhouse, just like in *Treasure Island*, to charge them revenue. Fishermen with an eye for adventure and quick profit turned into smugglers. Small vessels brought in casks of contraband rum, brandy, tea, silks and lace. All the locals were involved. Crabbers hid contraband in pots; tin miners moved it secretly through mineshafts. Landowners financed the operation and paid violent local thugs to fight off customs officers. The tax revenue paid for the British Empire, or whatever they called it when white men went and clobbered someone on the head who had been living there a long time. Pilchards fed the empire. One family fought pitched battles with the Crown's men-of-war. They raided Penzance's customs house to get their stolen tea back.

He looks out of the driftwood house. The grass is so long he can't see the stockade. He needs a sheep to keep it down, but can't afford one. What the fuck does he do for fun anyway? Walks his fucking dogs. Goes for a pint. That's the only fun he has. That's all he fucking does. Adults don't have fun like when you're kids.

Nutty formed a grand conspiracy. He linked Captain Avery and another famous pirate, Captain Kidd. He found Cadgwith closely resembled Black Hill Cove in *Treasure Island*. The Admiral Benbow Inn could be Cadgwith Cove Inn. His most controversial theory was that Treasure Island itself was in fact Kennack Sands. He found a description of Avery's men building defences on a beach with a ditch, a stockade and a line of cannon. He painted several canvases to show what this seventeenth-century garrison might look like, to help present his case. If he could find evidence of the garrison in the dunes, he could prove his theory. So he went out there and scoured around. He had a good idea where the garrison was. A cannon plunger had been found nearby. There were also three stones, hidden under grass, which corresponded to markers on a treasure map he'd found at the Cornwall Record Office. The treasure was beneath this spot. It made sense that the government had kept it secret. Avery's plundering of two Indian ships threatened to wreck trade with India.

He got in touch with a local archaeologist from the Cornish Archaeological Unit. They must excavate Kennack Sands and find the treasure. It could be worth up to £30 million. It would be worth a huge amount in extra tourism to the area. The archaeologist replied that he needed hard evidence of form and structure. Not historical records. He wanted to see a seventeenth-century garrison.

In 2012 Nutty Noah met two archaeologists on the beach. A report was made. They noted the presence of a marked stone. It was recorded in the online magazine *Rock Articles* in 2013. He kept chasing them up, saying he'd seen more evidence of the garrison. They told him to map it all out. Further meetings would be £262.50 a day.

Nutty Noah had friends. The Lizard people help each other, because it's so remote. Sir Ferrers Vyvyan, who owns the Trelowarren estate, will mysteriously find a bass, crab or cauliflower left at his door for favours he has done as high sheriff of Cornwall. The names of people on his farmsteads are the same as hundreds of years ago. One of Nutty's friends told him what he really needed was a radar.

Nutty explained to the lovely lady in the Cornwall Record Office that the Admiral Benbow in Chapel Street in Penzance cannot be the real one, because there is not a high cliff there, where Billy Bones sat, and you can't see boats on the beach. Robert Louis Stevenson holidayed in Mount's Bay for two weeks with his parents, before he wrote the book. Two artefacts had been found on the beach. They matched the design of Avery's garrison. Records showed a customs officer, tailing the pirate William Kidd, had searched the same beach. People told Nutty he had gone insane.

Nutty camped out for nights on Kennack Sands. He dug deep into the dunes. The police moved him on.

The local archaeologists pointed out that any treasure discovered is reportable under the Treasure Act 1996. If he finds one gold coin he has to take it to the district coroner within fourteen days.

When he pioneered ring-netting pilchards, people laughed at him. But he didn't give up. So he keeps sending questions to the Cornwall Record Office. An eminent Cornish historian told him he'd drunk too much Cornish mead. He feels he is taking on the world. If he finds the treasure will he take it to the coroner? The Cornish don't always take notice of English rules.

Sometimes when he's driving down the Lizard on his taxi job he sees a yacht going west. South of the Lizard. He drops his fare off, drives back and then sees the yacht again. It hasn't gone anywhere. It's still out there. For a moment Nutty finds himself on the boat. He can feel the tide against him. There's a westerly wind too. It's coming straight from the Atlantic. He's no longer sheltered by the headland. He's opening up a lot of sea there. The tiny white boat pushes forward against it. He might as well draw in a whale with a fishhook. The sea cares for nothing and no one.

The boat passes through the waves, and the water closes up behind, leaving no trace.

BIBLIOGRAPHY

There is a large collection of scholarly and consuming works on Cornwall. I only read a select list because the heart of this book is the voices of the characters and stories of living people. My primary sources were interviews with those I met. It was fascinating and insightful to read the following books that covered different aspects of Cornish life, culture and history.

Berlin, Sven, *The Dark Monarch*. Finishing Publications, 1962.

Button, Virginia, *St Ives Artists: A Companion*. Tate Publishing, 2009.

Causey, Andrew, *Peter Lanyon: Modernism and the Land*. Reaktion Books, 2006.

Du Maurier, Daphne, *Vanishing Cornwall: The Spirit and History of Cornwall*. Victor Gollancz, 1967.

Garlake, Margaret, *Peter Lanyon*. Tate Publishing, 1998.

Greenwood, Paul, *Once Aboard a Cornish Lugger*. Polperro Heritage Press, 2007.

Greenwood, Paul, *More Tales From a Cornish Lugger*. Polperro Heritage Press, 2011.

Halls, Monty, *The Fisherman's Apprentice*. AA Publishing, 2012.

Hepworth, Barbara, *A Pictorial Autobiography*. Tate Publishing, 1970.

Jones, Robert, *Alfred Wallis: Artist and Mariner*. Halsgrove, 2001.

Knight, Sam, 'A God More Powerful Than I'. *Harper's*, 2014.

Lawrence, D. H., *Kangaroo*. Penguin Group (Australia), 2009.

Marsden, Phillip, *Rising Ground: A Search for the Spirit of Place*. Granta, 2014.

Marsden, Phillip, *The Levelling Sea*. HarperPress, 2011.

McWilliams, John, *The Cornish Fishing Industry: An Illustrated History*. Amberley Publishing, 2014.

Miller, Amos C., *Sir Richard Grenville of The Civil War*. Phillimore, 1979.

Nowell, Roger, and Jeremy Mills, *The Skipper: A Fisherman's Tale*. BBC, 1993.

Payton, Philip, *Cornwall*. Alexander Associates, Fowey, 1996.

Payton, Philip, *Cornwall's History: An Introduction*. Tor Mark, 2002.

Pierce, Richard, *Pirates of Devon & Cornwall*. Shark Cornwall, 2010.

Sagar-Fenton, Michael, *Penlee: The Loss of a Lifeboat*. Bossiney Books, 1991.

Stoyle, Mark, *West Britons: Cornish Identities and the Early Modern British State*. University of Exeter Press, 2002.

Symons, Alison, *Tremedda Days: A View of Zennor, 1900–44*. Tabb House, 1992.

Val Baker, Denys, *Britain's Art Colony By the Sea*. George Ronald, 1959.

ACKNOWLEDGEMENTS

As I was writing this book an image of my Welsh father watching Wales beat England at rugby kept coming to mind. Like many Welshmen nothing seemed to stir him so much as defeating the 'old enemy'. At his funeral they remembered 'A passionate Welshman in the body of an English gentleman' with pocket handkerchief, suit and clean fingernails who knew everyone as he moved through the cloisters of Guy's Hospital in London where he worked as a physician all his adult life. There is much of this contradiction about the Cornish. They are generous, warm hosts to all the visitors who come each year but in them also stirs an ancient Celtic pride and defiance, separate from their powerful neighbour. I am very grateful to the many Cornish people I spoke to or met during my research.

I'm especially grateful to Martin 'Nutty Noah' Ellis for giving up so much of his time to chat to me. He explained about fishing so articulately and always with warmth and humour. Thanks, too, to Sally and Toots: going fishing with her dad was one of the great experiences of her life.

It was fascinating to sit in Penzance and listen to Andrew Lanyon talk not just about his father but about Cornwall with such scholarship and enthusiasm. Stacey was great Falstaffian company in the Swordfish, and kindly shared the details of his life with me. Nick Howell gave up his time to help me record his life as a fish merchant in Newlyn. He covered the subject with real erudition. Thanks to him and his wife Marie-Thérèse who hosted me at their place as they remembered times past. I was also lucky enough to interview Edwin Madron at length. He was one of the last sea-dogs and sadly died shortly after we met.

I met Jude Le Grice during a Six Nations Championship in the Star Inn in St Just and we had a good talk about his life and family. He was intrigued that the journalist Sam Knight (no relation!) had only just finished an article about him. I met up with Jude several times and found him happier with each meeting. Lyn Le Grice was incredibly kind to invite me over for drinks to talk about Jeremy's painting, never once mentioning her own considerable success as a designer. I included a passage about Jude's tragic episode because it represents one of the many ups and downs the Le Grice family lived through. It shows that Cornish defiance, one of the key themes of the book. Last time I saw Jude he was in good spirits and singing again.

Billy Stevenson was enormous fun in his living room. So outspoken, so knowledgeable, a key figure in Newlyn's fishing dynasty. Elizabeth Stevenson, too, was kind to talk to me quite candidly in her house about the family business and answer a wide range of questions with good grace. I'd like to thank Patch Harvey, the long-term coxswain of the Penlee lifeboat. Few people know about St Ives as well as Harding Laity and it was hugely engrossing to sit and chat to him in his house. He is a captivating story-teller. Erica at Spider Eye talked to me about their work in St Just which represents a positive future for young Cornish. Ben Weschke told me about his extraordinary parents and his own work with Spider Eye.

Ben and Jackie Gunn kindly met me first in the Swordfish and then at their flat on Fore Street in Newlyn. I spent enjoyable times in the Red Lion and Star talking to Larry Ratcliffe, Perry Withecombe and Vince Nowell.

I'm very grateful to Blair Todd for talking to me candidly about art in Cornwall and his own memories at the Newlyn Art Gallery. Henry Garfit showed me the Newlyn Art School and told me about his own personal quest to set it up. Martin Val Baker lent me some of his father Denys's work and gave me useful insights to that era in St Ives.

I'd also like to thank Mike 'Butts' Buttery, Johnny McFadden, landlord of the Star in St Just, Dennis Pascoe, 'Cod', Leon Pezzacks, Andrew 'Blewey' Blewett, Des Hannigan, David Barron, Mary Ann Blomfield, Fi Read, Andrew Cowan Dickie, Kenny Downing, Joe Crow, Mark Curtis, Mike Tressider, Pol Hodge, Sir Ferrers Vyvyan, Robert Matthewson, Roger Trevorrow, Marna Judson, Robert Jones, Lewis and Lauren at Callestick Ice Cream in St Ives, Ben from the Attic, Leah Churchman and her team in the Sloop and Jackie from the Red Lion.

Thank you to our friend Rachel Wylie for highlighting the idea of rural justice in Cornish communities. Michael Sagar-Fenton has written the definitive account of the Penlee disaster so I was very grateful that he not only allowed me to quote from it but agreed to chat at his home on the cliffs. Thanks to Jeremy Mills for allowing me to use a story from *The Skipper* and to Paul Greenwood for letting me use a piece from *More Tales from a Cornish Lugger* and for other insights he shared over the phone. Sam Knight very graciously allowed me to use the research from his powerful article about Jude Le Grice in *Harper's*. Thanks also to the Newlyn Archive.

I am very lucky to be represented by Will Francis, my agent at Janklow & Nesbit. Far better read and smarter than me, he nurtured this project from its earlier fragmented form, and gave invaluable guidance. I was very fortunate to have Becky Hardie at Chatto & Windus as my editor. She shored up the structure and narrative with an expert eye. The text benefitted enormously from her rigorous edit and memory. The final form owes a great deal to her. I'm grateful to David Milner for his sensitive copy-editing which saved me from embarrassing inaccuracies.

Finally to Anna who put up with me disappearing to do research, and who edited, encouraged and showed great insights throughout. I owe her more than I can say.